TIME TRAVEL

Also by John W. Macvey:

Interstellar Travel
Journey to Alpha Centauri
Space Weapons/Space Wars
Colonizing Other Worlds
Whispers from Space
Where Will We Go When the Sun Dies?
Alone in the Universe?

TIME TRAVEL

John W. Macvey

Scarborough House/*Publishers*

Scarborough House/*Publishers*
Chelsea, Michigan 48118

FIRST PUBLISHED IN 1990

Printed in the United States of America

Text design by Jennifer Lindsay

Final graphs by Basil Charles Wood

Library of Congress Cataloging-in-Publication Data

Macvey, John W.
Time travel: a guide to journeys in the fourth dimension / John W. Macvey.
p. cm.
Includes bibliographical references.
ISBN 0-8128-3107-1: $16.95
1. Time. 2. Time travel. I. Title.
QB209.M23 1990
530.1′1—dc20 89-78384
CIP

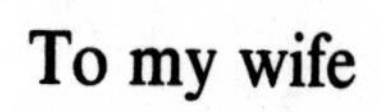

To my wife

Who can undo what time hath done?
Who can win back the wind?
Beckon lost music from a broken lute?
Renew the redness of a last year's rose?
Or dig the sunken sunset from the deep?

—Edward Bulwer-Lytton

Acknowledgments

My sincere thanks go out to those who helped bring this book to fruition: my wife and my younger daughter for invaluable secretarial services, and Helen Duncan, who was again responsible for converting my crude attempts at artistry into acceptable diagrams. I must also express my thanks and gratitude to Dr. John Gribbin, who readily granted me permission to reproduce two diagrams from his book *Time Warps.* These form the basis of figures 23 and 24. I am also indebted to Dr. Gribbin for some of the ideas in chapters VIII, IX and XI, and to the Open University, Milton Keynes, England for the basis of the calculation in Appendix 1.

Acknowledgments

[illegible]

Contents

Preface

When my publishers invited me to write a full-length work devoted to the subject of time travel, I viewed the prospect, I must confess, with some trepidation. Certainly on a number of occasions during the preceding few years the idea had crossed my mind, but on contemplating the many queer and irrational paradoxes that the subject must invoke I always relegated the intention to some indefinite future occasion. Now clearly that occasion had arrived.

I cannot claim that this book has proved easy to write. The nature of the theme effectively ensures that. Moreover, during the many months spent on research and planning there was one essential factor I dared not ignore or lose sight of—the text must be kept as straightforward and "popular" as possible. Many an involved concept necessitating several pages of explanation, and sometimes more repetition than I would have preferred, could have been disposed of by a single page or so of relativistic mathematics or abstruse detail, but the substance of a book embodying this sort of approach would hardly have been accessible to most lay readers. As it is, many of the fundamentals involving travel within the mysterious fourth dimension we identify as time seem a virtual affront to rational thinking and common sense. Nevertheless, the subject is an incredibly fascinating one with implications (and complications) bizarre in the extreme. As I gradually involved myself more deeply than heretofore in the subject, I found it had something of the property of a very potent drug, one on which I steadily became more and more "hooked." Whether or not the facility to travel through time, backward into the past or forward into the future, would be advantageous to humanity is a question I do not feel competent to answer with any assurance. Most readers, I suspect, will share my ambivalence. The power has not been given to us to foretell the future, and many would no doubt agree that this is just as well. Nevertheless, human beings are by nature questing, striving creatures, and if time travel ever can be rendered a feasible and practical proposition they almost certainly will wish to add this supreme technological breakthrough to an already imposing list of successful battles against the seemingly impossible.

In the interests of clarity I have quite deliberately simplified several diagrams. There exists a series of standard diagrams relating to travel within and through black holes that appear in many scientific papers, journals, and more erudite treatises on time travel. Thinking that these diagrams were too involved and did not easily present a mental picture of what could be happening, I decided that it might be preferable to portray a black hole (and forms of possible travel though it) as a black hole really appears—or is thought to. On only one occasion have I resorted to the standard type of diagram (in figure 24) and then only because there was no valid alternative.

I hope that in the pages to follow I have been able to shed a modicum of light on the weird realm of time travel, a phenomenon Lewis Carrol fans may be tempted to call "curiouser and curiouser!" In the bibliography I have listed a considerable number of papers, articles, and books. Some of these represent reasonably straightforward going, but many are fairly abstract, calling in some cases for the knowledge and ability to cope with rather involved mathematics. In those instances the going may prove hard but the reward should be worth the effort.

John W. Macvey

TIME TRAVEL

INTRODUCTION

Time's but our playmate, whose toys are divine.

—Thomas Wentworth Higginson

What, precisely, is time? At the outset we can admit that our understanding of it is anything but "precise." But since we must attach a handle to the concept, is it a particular moment; a period of duration; the conception of past, present, and future as a sequence; the limited duration of human existence as contrasted with eternity? None of these dictionary replies is satisfying. Time is then the awesome name of the fourth dimension, "something" that seems to flee from definition in its fullest sense.

We are all well aware of time—or more correctly, we are all *too* well aware of its passing. For the individual it starts the moment he or she is conceived or draws his or her first breath, and apparently it ends when the last breath leaves the body. Like the other common dimensions, we can measure it, though not quite in the way we are able to measure length, breadth, and depth. We know it exists because if it did not, neither would we or the universe around us. But it is not something tangible in the orthodox sense. We cannot handle it; neither can we look at it. If we glance at a wristwatch or clock, we can see that it is passing—and with every second measured our respective lives are brought that much closer to their inevitable end. On our faces and bodies the results of a few minutes, hours, or days do not show, but if we choose to compare recent photographs of ourselves or friends with snapshots taken ten or twenty years ago, the effect of time's passing is all too clear. In this context it is perhaps hardly surprising that we often refer to time by its "ravages." It is, then, very natural to wish that somehow we could reverse the seemingly inevitable flow of time, to jump back all those years. But of course we cannot. We may only mourn our lost youth. And how many of us truly would desire to jump ahead in time were this a practical proposition? It would prove interesting and highly instructive to see how the progress of history had gone. But—and here is the first of the many paradoxes—would we age as we went or stay just as we are at present and find ourselves gazing ruefully at our own tombstones? If that proved depressing (and presumably it would) and we could then return to the present, we would remain depressed since now we would be well aware of the precise date of our eventual demise! (As once said of hanging, it would concentrate the mind most wonderfully.) Moreover, a trip into the future followed by a return to the present could conceivably enable us to influence the future course of history in some way. But if that's so, the future we had beheld could not then have been the real future! In Chapter VI ("Causality and Paradox") I have endeavored, so far as is possible, to explain these paradoxes and suggest ways in which they might be avoided. I have not been able, however, to answer categorically whether or not time travel can ever really be a practical or a desirable thing. It would be a very sanguine person indeed who would care to conclude one way or the other. All we can do is to investigate and examine the strange medium and try to come to grips with the implications of travel within it. It is only fair to place on record, however, that a number of physicists are coming around to the idea that travel in time *is,* theoretically at least, a feasible proposition, though one hedged by many restrictions. The concept of a time ship or time machine that we could enter at will and so transport ourselves back into the past or forward into the future by the manipulation of a few weird controls and gadgets should probably (and preferably) be seen as something belonging essentially to the realm of science fiction. Science fiction is nevertheless interesting because here at least is an intellectual setting in which time travel can be given full sway, though often one is compelled to admit that many of the awkward and paradoxical implications that so frequently arise are conveniently ignored by the genre writers or are glossed over with a veneer of pseudo-science. A classic example I came upon recently involved a time traveler belonging to a future era on Earth who elected to visit this present day and age. If that were possible, it is perfectly reasonable to assume that some of us would see or even meet this intrepid voyager in time. Of course for obvious reasons he might not be too keen to admit whence he had come. Should he elect to dispose of his futuristic raiment and don the garb of the 1990s none of us presumably would need be any the wiser. Indeed we might find ourselves addressing a great-great-grandson not yet born! As I have said, the fantasy writers generally gloss over such dilemmas. Another recent short story proved to be one in which the author tackled the theme in an imaginative way—UFOs, flying saucers, and all other examples of space-borne crockery do not come from outer space, it seems. They are *time ships* peopled by our distant descendants from a far future period on Earth! Should, by any chance, this really be so, they seem (perhaps understandably) reluctant to come face to face with their ancestors.

One of the first persons to introduce the concept of travel in time was the celebrated British novelist H. G. Wells*; his famous novel *The Time Machine* was written in 1895, just before the turn of the century. It featured a vehicle capable of bearing its occupant into either the past or the future at will. Since Wells wrote *The Time Machine* a not insignificant proportion of science fiction has been devoted to this highly intriguing theme, although space travel has always enjoyed precedence, perhaps because it is so much easier to explain and its near-term prospects so much more demonstrable.

In a 1975 article, science-fiction writer Harry Harrison advanced an idea that I intend to dwell on more fully later in these pages. His idea was expressed as follows: "What if time is more like an ever-branching tree with countless possible futures? If each decision we make affects the

---

*Wells, of course, would have been familiar with the trips to "Christmas Past" and "Christmas Future" taken by Charles Dickens's character Ebenezer Scrooge in *A Christmas Carol*, published in 1843.

future then there must be an *infinite number* of futures. In the river-of-time concept the future is immutable. If on the way to work in the morning we decide to take the bus instead of the 'tube' (subway) and are killed in a bus accident, then that death was predestined. But if time is ever-branching then there are two futures—one in which we die as a result of the accident and another where we live on having taken the 'tube.'" More about this in Chapter XI ("Other Universes—Sideways into Time").

We can surely see in this a totally new concept of time. To use a different analogy, it is no longer the ever-rolling river that carried us from its headwaters (our birth) down through its midcourses (youth and middle age) to the broad estuary where it meets the sea (death). Time can be thought of as a stream that branches out in a number of different directions to form entirely independent rivers all of which must eventually reach the ocean but at different points on the coast.

I asserted earlier that if the dimension of time did not exist, neither would we, nor would the entire universe surrounding us. Before me on my desk as I write is a small slab of marble. It possesses length, breadth, and depth, the three normal dimensions of which we are so well aware. But if that mystical fourth dimension, time, were not present, that block could not possibly exist. It requires time in which to exist just as do the farthest stars and galaxies. Thus it can be seen that space and time must be inextricably related. There cannot be one without the other, which is why we really should refer to space as the space-time continuum or as space-time. Later I'll be going more fully into this relationship and the very odd implications that accrue.

Clearly if we are seriously to consider the possibilities of travel in the fourth dimension, we must try to understand more about the nature of time itself. If matter can be compressed or expanded, can time be treated in the same manner? To appreciate time in this way we first must take a close look at space in a totally different way. To do this we must consider space as more than just an infinity of "nothingness" in which stars, planets, moons, and galaxies plow their lonely and apparently eternal paths.

This, then, will be one of our first objectives—to look at the universe as it is, not as we may think it is. The universe is not just queer. It is a whole lot queerer than we really imagine. We will look first at the supposedly conventional universe of Isaac Newton and then at the much stranger universe of Albert Einstein. Before embarking on this preliminary investigation, however, it is necessary to devote a chapter to what we might conveniently term the history of time, its effects on our everyday lives, and its measurement through the ages. At first this may seem to have no apparent direct connection with time travel, but it's a necessary grounding from which we can build. Every structure of consequence must have its foundations. So also must any solid concept of something as amorphous as time.

2:00

TIME

In one of his celebrated essays Charles Lamb wrote, "'What is truth?' said jesting Pilate and would not stay for an answer." We may equally well ask again, "What is time?" and no matter how long we stay it is very doubtful if we will find a really satisfactory answer. You'll recall the standard dictionary definitions I quoted in the preceding chapter, but it can hardly be said that any of them is really the answer we seek.

In a very broad sense we are already travelers in time. We started on our journey at the moment of conception and that journey will end at the moment of our death. But this journey of ours is a thoroughly unremarkable one into the future at a pace over which we have no control whatsoever. Moreover, we cannot reverse the direction of our journey. Tomorrow will come, but yesterday is irrevocably gone. It could be said that virtually every action we take in one way or another influences our future and perhaps also that of others, just as theirs may have a direct bearing on our own. It is something over which we have no control. We can move more or less freely in the normal three dimensions, but in that strangely insubstantial fourth one, time, it seems we are as helpless as a fly caught in a spider's web.

To our very early ancestors time could have been little more than a mere sensation—even the notion of trying to describe it would have been beyond them. As civilization developed, it became increasingly essential that time, or more correctly the rate of its passing, be measured. At first this posed a considerable problem. Eventually progress was made by recourse to observations of the motions of the heavenly bodies. It is virtually certain that early "almanacs" were based on these motions, notably those of the Sun and the Moon. Those people could hardly have failed to observe that the Sun rose and set at slightly different times each day and that this paralleled what eventually came to be known as the seasons. They would have been aware, just as we are today, that as winter in all its harshness gave way to spring, not only did vegetation again come to life but the Sun rose slightly earlier in the east each morning. The early astronomers would also have noted that as the season progressed, the Sun also rose from a different point on the eastern horizon: that in what we call summer it rode high in the sky whereas in winter its daily path across the sky from east to west was considerably lower. The ancients must also have observed that the Moon acted in a converse fashion, more recently giving rise to that well-known adage "the Moon rides high in winter and low in summer."

The lunar cycle or lunar month would also have been apparent. Our early ancesters would have seen, just as we do, the "new moon" as a very thin crescent hanging in the western sky at sunset. Each succeeding night, given clear skies, they would have noticed two things—that a greater proportion of the Moon's disc was visible and that its position in the sky indicated that it must have risen slightly later. After "full moon" they noticed how on each ensuing night less and less of the disc was visible until ultimately, just prior to sunrise, the Moon hung as a thin crescent in the eastern sky. Then of course it was the turn of the opposite "limb" or rim to be illuminated.

So ancient peoples had their "day" depending on the length of the shadow cast by the Sun on primitive sundials—but only of course during daylight hours and under clear skies. They had their month thanks to the lunar cycle. But what of the period of 365¼ days, Earth's orbital period around the Sun, that we call a year?

Ancient civilizations, especially those in the Near East and along the north African shores, were in many cases nomadic; on clear nights, unplagued as we are by the glare of street lights, neon signs, and smog, they would have seen, especially on moonless nights, the whole glory and panoply of the star-powdered heavens. They must have noticed that the stars, as did the Sun, rose in the east and set in the west, and as the months passed, that each star and constellation rose earlier and set earlier to be followed with seeming endlessness by the appearance above the eastern horizon of fresh stars and constellations. The constellations of spring differed from those of winter, those of summer from those of autumn, those of autumn from those of winter. It was a steady celestial round.

To sum up, an observer in the northern hemisphere viewing the stars at the same time on each successive evening while the Earth makes a complete orbit of the Sun will notice that the positions of the stars move gradually in a counterclockwise direction around the North Star (Polaris). After the passage of one full year the stars and constellations again appear in their original positions. In this way ancient peoples could tell that a year had passed.

Many of these societies were largely dependent on agriculture. There was, for example, a time for plowing, a time for sowing, and a time for reaping. They required a calendar or almanac to indicate with some precision the beginning and end of the various seasons; the heavens, in their unique and unfailing way, provided just that. To us in this age of quartz and atomic chronometers accurate to the minutest fraction of a second, time measured by sundials, the phases of the Moon, and the annual rotation of the stars may seem almost barbaric in its primitiveness. But it was a start. Indeed, were it not for methods such as these, several very ancient and valued architectural artifacts would not exist; as we now know, many of these objects were created to serve as a form of clock, calendar, or almanac. There are various monuments or artifacts of this kind around the world. Probably the best-known example is located on Salisbury Plain in the county of Wiltshire, England, and is known as Stonehenge. It is essentially a group of standing stones of Neolithic age, believed to have been erected in the second millennium B.C. It comprises four series of giant stones encircled by a ditch 300 feet in diameter. The outermost series is composed of sandstone, the megaliths being arranged in a circle. This encloses a circle of "bluestone" menhirs (very tall and massive stones). The third series is shaped like a horseshoe. The inner ring has an ovoid configuration with an altar stone within its confines. A huge upright "heelstone" is positioned to the northeast of the circle.

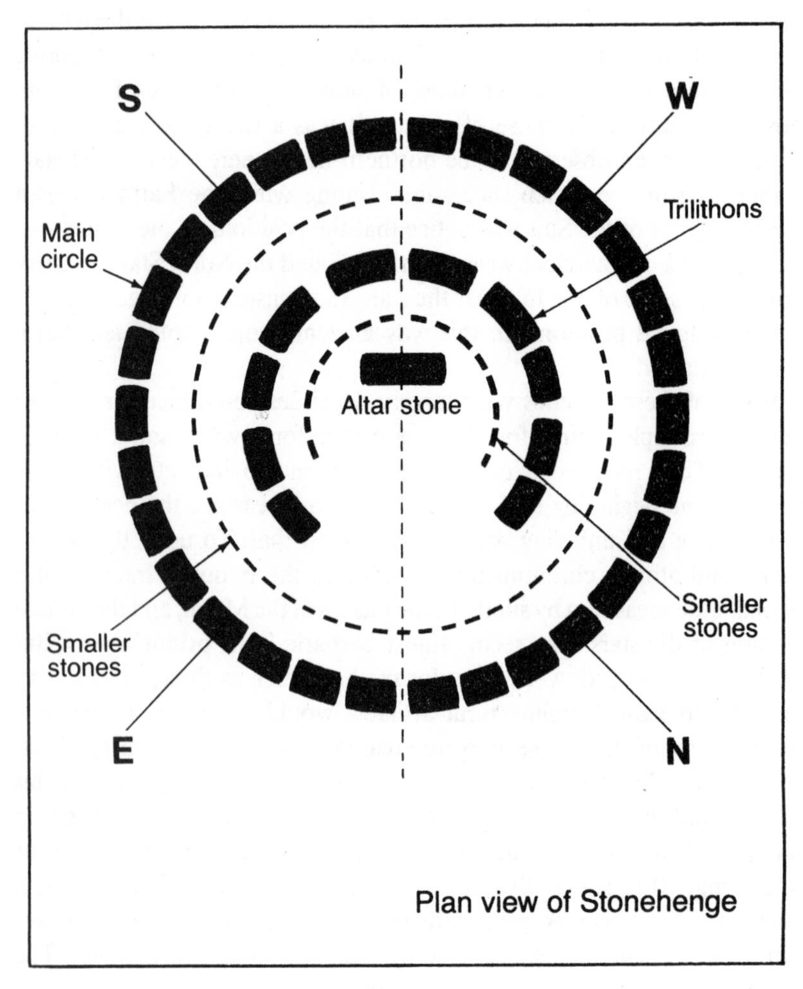

Figure 1

Sketch plan of Stonehenge in its original state.

A long-held and widely-believed theory that Stonehenge was once a Druid temple is now discredited. Druids were priests of the ancient Celtic peoples of Britain and Gaul who, for some obscure reason, worshipped the oak tree and probably also the Sun, but Stonehenge predates them by a very considerable period.

It was only in the 1960s that the hypothesis that Stonehenge had been built primarily as an astronomical "instrument" gained broad credence and acceptance. It probably also had a pagan religious significance as the presence of the altar stone would seem to testify. Credit for this discovery must be attributed to archaeologist Gerald Hawkins, who employed a computer to help assess the significance of the various stone alignments. It is worth recording that noted British astronomer and cosmologist Fred Hoyle carried out a program of related research without any electronic aids and reached the same conclusions.

Figure 1 gives a fairly accurate representation of the plan view of Stonehenge as it must have appeared when its construction was completed. In the outer circle were thirty large stones. Each pair of stones was capped with a lintel or large horizontal stone. The inner horseshoe of stones or "trilithons" also were arranged in pairs of uprights and with a crosspiece. The altar stone was positioned roughly in the center.

In the course of early archaeological study, it was soon noticed that the axis of the stone circles lay on a line running through the center of the altar stone to another stone beyond known as the "Friar's Heel" and that this line must have pointed directly toward the point of sunrise, at the summer solstice, when Stonehenge was completed. It is an established fact that the daily point of sunrise varies slightly from year to year owing to changes in the Earth's course around the Sun. The celebrated British astronomer Sir Norman Lockyer once calculated that this line pointed directly toward the point of sunrise at the summer solstice in the year 1680 B.C., thus providing the probable date of completion of Stonehenge. Recent investigations suggest, however, that Stonehenge was not constructed all at once, but in three entirely disparate phases from 2600 B.C. to around 1700 B.C.

If you visit Stonehenge at dawn on "midsummer's day" (June 21) of any year and position yourself at the very center of the edifice,* you will observe that the Sun rises exactly over the "heelstone." On no other day of the year will this take place. It is just one of the more obvious manifestations of time's annual rhythm. Moreover, though somewhat less obtrusively, certain other stones on the site, stones laid out in the configuration of a rectangle, provide what we might conveniently term "lunar markers" since they serve to indicate the most southerly rising position and the most northerly setting position of the Moon over a cycle of 18⅔ years. You may be inclined to ponder the precise significance of this seemingly arbitrary figure. In fact it is far from arbitrary, as we will see. The Moon, unlike the Sun, does not manifest an exact annual cycle with respect to its rising and setting. I have already mentioned how the Sun, during the course of a year, appears to rise from and set over a different part of the horizon. Put in more precise terms, the Sun's regular cycle in this respect is one in which it rises a little farther to the north from the first day of winter to the first day of summer, and thereafter that much farther to the south until the winter solstice is again reached. The Moon, on the other hand, rises more toward the north on certain days, more toward the south on others. There is a definite cycle to this that is repeated every 18⅔ years. Since deduction of this far-from-obvious fact could only have resulted from meticulous observation spanning several decades before the correct alignment of the appropriate stones could be achieved, our feelings toward these very early astronomers and timekeepers (for that is what they were) must be respectful.

But now we must return to the basics of time measurement—how through history human beings have measured the inexorable one-way flow of the mysterious fourth dimension. Passing from our look at one manifestation of Neolithic civilization in northern Europe, we see that

*Entrance to Stonehenge *on June 21st* has been prohibited since that day in 1988, when a large number of overzealous visitors, described by one observer as "hippies and drug addicts," reportedly abused the place and engaged in a running battle with the police.

ensuing centuries led to other, more developed, civilizations along the shores of the Mediterranean, in what is now called the Middle East or Near East, and eventually to the rise and subsequent fall of the mighty Roman Empire. Thereafter, for reasons most schoolchildren know, Europe degenerated into what came to be known as the Dark Ages. Through these dismal centuries of ignorance, superstition, and frequent plague, the measurement of time with any degree of accuracy must have had a very low priority indeed. It was, however, taken up by others in lands far from Europe. It is probably the Chinese to whom most credit must go for their achievements during that epoch; for they were responsible for the invention of timekeeping instruments that can reasonably be regarded as the direct forerunners of modern clocks. Their "clocks" were activated by water power, and the motion of these devices was arranged to coincide with the apparent daily rotation of the stars.

The earliest mechanical clocks to have movable parts were in all probability produced around 700 years ago, although the first instrument to measure daily time seems to have made its appearance more than 3,000 years ago. This was the Egyptian "shadow" clock. A rather primitive sundial, it measured daily time (during daylight hours) by the movement of a shadow across fixed markers. This was followed by a water clock known as a clepsydra, although it differed markedly from the Chinese water clock mentioned in the preceding paragraph. Water was supplied to a funnel, which channeled it to a cylinder containing a float. The rising float was coupled to a rack and pinion device actuating a single hour hand. The rate of water flow was regulated by a graduated stopper, the water being maintained at a constant level by means of an overflow tube. The sand glass or hourglass also first appeared around this period, time being measured by the change in level of flowing sand or water.

Most inquiry indicates that these devices remained the only methods for measuring daily time until the Anglo-Saxons introduced the use of candles marked at regular intervals. During medieval times instruments had dials calibrated only in hours. The advent of minute (and second) measurement was still some way off. These included not only the fully developed sundial but also the star dial. It's obvious that a sundial will function only when the Sun is above the horizon, but the necessity to measure as well the nocturnal hours became the mother of invention. Thus the star dial, appropriately termed a "nocturnal," was devised. This was composed of a dial, calibrated in hours, to the center of which was pivoted a pointer. The North Star was sighted through a central hole in the dial, the pointer being rotated toward the two "pointer" stars of the Big Dipper. On cloudy nights, alas, time measurement remained a matter of guesswork.

Falling weights as driving mechanisms suffered from one rather notable disadvantage—they were far from portable. Thus by the fifteenth century we begin to see the genesis of compact clocks deriving power from metal springs. Though certainly portable, these early spring-powered clocks were far from accurate, especially when the tension of the spring began to fail; in those days that was soon. Eventually a minute hand made its appearance, but a hand recording seconds was virtually unknown until the advent of the pendulum. The pendulum was the invention of a seventeenth-century Dutch scientist, Christian Huygens, although the underlying idea was almost certainly Galileo's. With the arrival of the pendulum, clocks at once began to achieve a much higher degree of accuracy. But problems of accuracy still existed mainly because pendulums were affected by changes in temperature which, by virtue of expansion and contraction, led to changes in their length and thus to slight changes in their period of swing. Not until 1715 was this particular problem overcome when George Graham devised a technique to compensate for temperature changes.

In 1675 Huygens introduced the balance or hairspring, which, like the pendulum, led to greater accuracy. Like the pendulum, however, it was adversely affected by variations in temperature. Despite this shortcoming, Huygens incorporated it into a "watch" designed for the determination of longitude at sea. It was not until 1753 that an effective method of compensation was devised, by John Harrison, resulting in a watch that is reputed to have erred by only five seconds over a six-week period.

Development and refinement of watches continued with the years. In the mid-nineteenth century a device known as the lever-escapement became virtually standard, enabling an even higher degree of accuracy to be achieved.

Clearly we have come a very long way from Stonehenge, the sundial, and the graduated candle, but the past few decades brought about a more profound revolution in measuring the rate at which time passes. Other types of clock, including those that employ an electric motor to wind a spring, made their appearance and seriously rivaled purely mechanical clocks. Electric clocks with synchronous motors have achieved a very high degree of accuracy because the motors are in phase with the frequency of the alternating current power supply.

The most recent achievements are quartz and atomic clocks. In the former the quartz crystal vibrates when the appropriate alternating voltage is applied across it (the piezoelectric effect). Such clocks are capable of an accuracy of a tenth of a second per year. Cesium-beam atomic clocks give even greater accuracy, responding as they do to oscillating energy changes within atoms.

It is appropriate to say a little on the subject of *time zones,* the phenomenon that so tends to upset long-distance travelers by modern jets. Periodically I travel between Britain and the United States and can speak from experience on what this form of transit (dare we call it "pseudo time travel"?) does to one's personal metabolic clock. Flying west across the Atlantic, one is in effect "following the Sun." On a normal five- to six-hour flight the day appears to be almost unending. It is as if the Sun were more or less stationary in the sky. One arrives in the United States at, say, 6 P.M. EST, but the body is quite assertive that it is 11 P.M. GMT and time for bed. When 11 P.M. EST comes around, it is 4 A.M. GMT, and the need for sleep has become very apparent. Traveling west to east across the Atlantic tends to be even more traumatic. As the flight toward Europe proceeds, dawn is met somewhere over mid-Atlantic. One has caught up with tomorrow! This does not, of course, constitute true time travel. Whatever hour the respective clocks are registering, people in

London, New York, San Francisco, Tokyo, or New Delhi are living at the *same* instant in time.

All this must be attributed not to some weird relativistic effect but to our planet's normal daily 24-hour rotation—the reason why the Sun appears to rise in the east and set in the west. Thus in western Europe during summer, sunset and the growing darkness of approaching night means that it is late afternoon on the eastern seaboard of the United States and early afternoon on the west coast. The terminator—a diffuse strip that separates day from night on our planet—sweeps around the Earth continually. Thus every line of longitude on the surface of the Earth must, strictly speaking, have its own particular time with respect to the Sun. However, to have time zones based on individual lines of longitude would be as ridiculous as it would be thoroughly impractical. For this reason a certain number of lines of longitude are grouped together to constitute a definite number of so-called time zones. Were this not so, relatively short journeys east or west would necessitate several clock adjustments, while on long journeys we would hardly ever stop making these adjustments. In fact, it was not until 1883 that the United States adopted a system of standard time by which the country was divided into four zones, the clocks within a whole zone reading the same time.

Zones between lines of longitude are 15 degrees wide and there are 24 of them. This is no arbitrary choice. Every 24 hours the Earth rotates through 360 degrees, i.e., any point on its surface makes a complete circle (with the exception, of course, of the poles), and dividing 360 by 24 gives 15 degrees. So when it is 12 noon at any point on the surface of our world, it will be 11 A.M. at a point 15 degrees to the west and 1 P.M. at a point 15 degrees to the east. (There are some slight variations in time-zone boundaries to accommodate geopolitical anomalies.) Traveling across all 24 of these time zones could produce a strange but entirely rational effect. Suppose, for example, that you flew nonstop from *west to east* around the world. Your watch would require to be advanced by 24 hours—or, more correctly, your calendar would—for you would *appear* to have gained a whole day. A similar journey this time from *east to west* would require that your watch (or calendar) be set back by 24 hours, resulting in the seeming loss of a day—more pseudo time travel. Again we must emphasize that you would not in the true "time-travel" sense have advanced either a day into the future or one into the past.

Let us examine the position a little more closely. You (at approximately 42 degrees north latitude) are proceeding nonstop by jet westward at a speed of around 700 mph. You will hardly fail to notice that the Sun shows no apparent change in its position in the sky. It simply seems to hang there immobile. Whatever time you left your starting point, it will still be that time when the aircraft touches down at the airport from which it departed. You will have "followed the Sun" precisely. The vital question will of course be Which day is it? To those who watched the aircraft depart it will clearly be the following day. In your mind, though, there must be considerable confusion, for if the time of departure was noon, you have experienced a noon that has lasted for 24 hours with no dusk, night, or dawn at all. What the human metabolic "clock" would make of all that is hard to say. A certain degree of disorientation compounded by lack of sleep could certainly be expected. It is apparent therefore that somewhere along the way you would find it desirable to advance the calendar by one day to remain in phase with friends left behind. At which point would this prove most convenient? The most logical would be at a point halfway around the world from the Greenwich zero meridian, i.e., at a longitude of 180 degrees. Fortuitously this particular meridian runs for the most part through the great trackless wastes of the Pacific Ocean. This eliminates the need for part of one land mass to be in a different (calendar) day from another with all the complications this would create.

Of course the presence of the International Date Line inevitably leads to a few perceptual anomalies. Travelers crossing from east to west must regard themselves as having gone from "today" into "tomorrow." If the direction of travel is from west to east, the converse results: the travelers have gone from "today" into "yesterday," though *not* of course, I must stress, in actual fact. Time travel, as we shall be seeing, is not that easy.

It is perfectly feasible for one to "repeat" a day *only so far as the calendar is concerned.* No matter how pleasant yesterday was, we can't really repeat it—at least not this way.

Once or twice I have referred briefly to the effect of jet travel on the body's metabolism—your natural inbuilt "clock." It might therefore be pertinent to close this chapter with a short discussion of this aspect.

On a planet that has been turning steadily on its axis for countless millions of years, it is hardly surprising that the bodies of living creatures have developed a kind of natural rhythm. This manifests itself in sundry ways. For example, some creatures such as the owl are largely nocturnal. They prey by night and sleep by day. Even the domestic cat continues to show something of this tendency. (I should know—I have four of them!) Reproduction among living things relates largely to seasonal cycles, though differing with respect to various species. The human female menstrual cycle bears a marked resemblance to the monthly lunar cycle, although it is also influenced by physiological factors. This is by way of saying that living things on Earth, flora as well as fauna, have become accustomed to the sequence of day and night, of the months and of the seasons, and as a consequence show a distinct relationship and affinity to these. We and all other living organisms contain a biological clock, and just as a normal clock tends to go wrong if subjected to unusual treatment, so do our biological counterparts. Unusual treatment in this context certainly includes jet travel over long distances during which the natural rhythm of the Earth's rotation with which our bodies are in phase is being disturbed. As might be expected, the effect varies from person to person. Some are able to shrug off "jet lag" after a few hours whereas others are still feeling the effects after several days during which their bodily functions, such as eating and sleeping become reattuned. We are able to adjust a watch or clock in a matter of seconds. The metabolic clock adjusts automatically, but this process takes appreciably longer, for it must recalibrate itself with respect to day and night. At this point it's legitimate to wonder how this adjustment will take place if genuine time travel ever becomes a feasible proposition.

Now we must start to look at time in the totally new and bizarre way of traveling through it into the future or into the past. This is an entirely different concept. The time we have been considering so far is a seemingly slow-moving, ever-onward stream that is bearing us with inevitability down the course of our lives. When that stream reaches its ocean, we will die—and there is not a thing we can do about it. Some of us are still in the sparkling headwaters of that stream, others have already reached its broader mature channels. Still others are smelling the sea and beginning to feel the first currents of the estuary. So in a quite involuntary way all of us are time travelers, but only into the future and at a rate over which we have absolutely no control. However much we might like to reverse the process and find ourselves back up in the sunlit headwaters of our youth, this is not possible. Tomorrow and the day after (we hope) will come; yesterday and all the days that preceded it up to our birth are gone beyond recall. Could all this be changed? This is the great question to which we must now address ourselves, and to do so we must first try to comprehend the nature and complexity of the universe around us.

3:00

NEWTON AND HIS UNIVERSE

When one starts to take a serious interest in astronomy, it is most unlikely that he or she, at that stage, will be greatly concerned with such cosmological complexities as the shape of the universe or whether it is finite or infinite. These more abstract and philosophical concerns come later—often much later. This was certainly the case for me. In my early career I was perfectly content to accept space as simply space—sheer unending nothingness in which the Sun, Moon, planets, asteroids, comets, stars, and incredibly remote galaxies occupied their allotted positions and courses. That was that. All very neat and simple. To the average person with only a modicum of interest in astronomy, this, in all probability, is still satisfactory. But as one takes an increasing interest in the great immensity surrounding our small and undistinguished planet, certain other factors begin to emerge, and difficult questions start to suggest themselves. There comes an increasing awareness that space is not merely space, that perhaps the visible universe is not all that it seems.

Early astronomers were greatly preoccupied with the nature of the force that kept the heavenly bodies in their respective positions and paths. Clearly something did. The big question was what. Galileo (1564-1642) and Tycho Brahe (1546-1601) were deeply involved in an attempt to find the answer, but it was Johannes Kepler (1571-1630) who first expressed the position in pure mathematics. Tycho Brahe had for long been a scrupulously careful and tireless observer of the celestial scene, and indeed most of his life was devoted to this work. Fortuitously, Kepler had free access to all the observations of Tycho Brahe, and from these he was able to formulate his three famous laws of planetary motion—laws that have stood the test of time and hold good to this day.

First Law:
The planets all revolve in ellipses—not circles—with the Sun at one focus of the ellipse. This law is more or less self-explanatory.

Second Law:
The planets all sweep out equal areas with respect to the Sun in equal times. This law perhaps calls for a little more thought. What it really tells us is that planets travel more slowly in the part of their orbit that lies farthest from the Sun (Figure 2).

Third Law:
The square of the total time of revolution of a planet is directly proportional to the cube of its distance from the Sun. This law is a little more involved too. What it says is that the more distant planets move more slowly in their orbits than do the nearer ones.

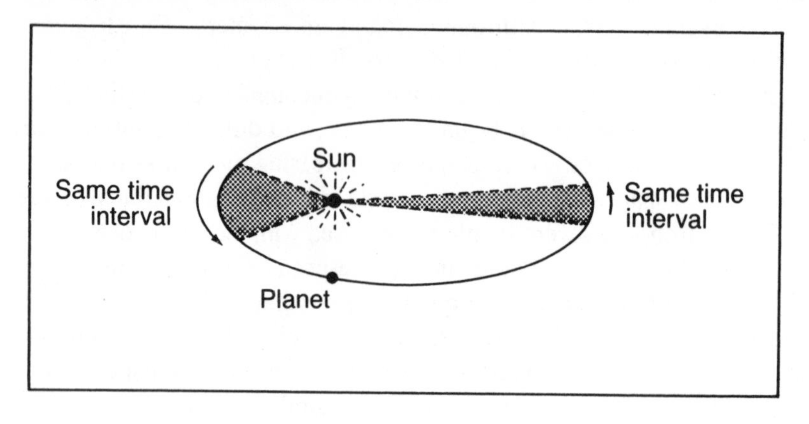

Figure 2

Equal areas swept out in equal times.

What has all this to do with the sought-after force or forces? From Kepler's second and third laws it is apparent that the farther a planet lies from the Sun the *less* is the force moving it in orbit. Whereas Kepler was convinced that the *source* of this motion must be the Sun, he found it difficult to reconcile the (focal) position of the Sun with the *elliptical* motion of the planets. The answer lay in an understanding of what today we term centripetal force. The nature of this force was unknown to Kepler, who thus postulated along different lines: "The Sun," he said, "remains in its place but rotates as if upon a lathe and sends out from itself into the depths of the universe an immaterial species of its body analogous to the immaterial species of its light. This species turns with the rotation of the Sun after the manner of a most rapid whirlpool or vortex throughout the whole extent of the universe and it bears the planets along with it in a circle with a stronger or a weaker thrust according as, by the law of its emanation, it is denser or rarer."

When Kepler wrote this he was going out on a limb to some extent, since at that time the fact that the Sun rotated was not known for sure. It remained for Galileo with his early refracting telescope and observations of the movement of sunspots around the Sun's disc to reveal that the Sun was indeed a body in steady rotation. Not surprisingly this revelation pleased Kepler greatly. In a subsequent letter to Galileo he wrote, "Under your guidance I now recognize that the celestial substance is incredibly tenuous. A single fragment of the lens interposes much more matter between the eye and the object viewed than does the entire vast region of the œther. Hence we must virtually concede, it seems, that the whole immensity of space is a vacuum."

This is hardly the way we would put it today, and certainly all references to the so-called "œther" would be deleted. Nonetheless a true understanding of the real nature of space was becoming more apparent. Kepler was, however, no nearer an answer as to why the planets should pursue *elliptical* orbits, and he continued to speculate on this theme. At one point he endeavored to equate the responsible force with light. In fact he had already derived a law that showed how the intensity of illumination decreased with the square of the distance from its source. This, it seemed, was analogous to the force acting on the planets. It was soon recognized, though, that light could not be the force. It was far too weak.

That force, as we all know today, is gravity, but in making this statement we are running well ahead of our narrative. Strange as it may seem to us, gravity was *not* then held responsible, being regarded in Kepler's time as a purely *terrestrial* force. If an object were picked up and then released, it immediately fell to the ground, the "terrestrial" force known as gravity being the cause. From his notes it is obvious that Kepler was well aware that any two bodies would attract each other and move together, although one or the other must be very massive for the effect to be apparent. A child's ball will fall to the ground because the Earth attracts it. The Earth is also attracted by the child's ball but only to an infinitesimally small extent.

At this point it's appropriate to define gravity more precisely. It is the name assigned to that force of nature which manifests itself as a mutual attraction between masses; its mathematical expression was first given by Newton (1642-1727). The law states that any two particles of matter attract one another with a force directly proportional to the product of their masses and inversely proportional to the square of the distance between them. This may be expressed by the equation:

$$F = G\frac{m_1 m_2}{d^2}$$

where F is the force of gravitational attraction between bodies of mass m_1 and m_2 separated by distance d. G is a term known as the constant of gravitation.

If we seem to be attaching a high degree of significance to gravity, this is because it is fundamental to the space-time continuum and the bodies therein. Moreover, as we will shortly see, gravitational fields of high intensity could have a very considerable bearing on our theme, notably with respect to travel into the past via black holes.

Kepler's notes also indicate that he thought gravity responsible for the

Moon-produced tides in the seas and oceans of our planet. Nevertheless, he could not, from all accounts, accept the premise that gravity might affect the entire Solar System. Just why he shied from that conclusion is none too clear, but it appears that he envisaged some less mundane and familiar a force, something more in keeping with the majesty of the universe.

Another force accorded serious consideration was, perhaps not unnaturally, magnetism. This had been studied as far back as 1600 by the English physicist William Gilbert. Kepler gave a great deal of thought to its possible implications. Could this be the force that bound the universe together and kept the planets orbiting the Sun in their elliptical orbits? Gilbert had already demonstrated that the force of magnetism would operate in an artificial vacuum and therefore, by definition, in natural space also.

It is not surprising that Kepler was eager to adapt the concept of magnetism; he immediately cited it to "explain" the alternate acceleration and deceleration of the planets as they moved around the Sun. That it proved a false dawn does not surprise us today. Kepler's use of the notion even then was somewhat less than convincing, but it did at least eliminate the need to postulate a "mysterious indefinable 'œther'" and, since it involved only the Sun's attraction, enabled Kepler's "vortex" theory to be dispensed with.

Kepler had, despite all his efforts, failed to identify the true nature of the force so fundamental to the universe. We must now turn to a man whose name has become legendary—Isaac Newton. Newton blew away many old shibboleths and fancies that had been impeding the progress of physics. Many of them were a legacy from the time of Aristotle and should have long since been abandoned.

Newton was faced with three distinct problems. He had to determine the *nature* of the force existing between the Sun and the planets, the form and function of the space separating them, and whether or not the known movement of the planets could be deduced with reference to this force.

Newton soon decided that circular movement was inherently unnatural.

He further maintained that constant velocity in a *straight* line epitomized the normal passage of an uninhibited body anywhere in space, motion that would continue in the absence of any external force. If a change did result, it was because an external force was acting upon that body. This might be represented by a change in velocity or a change in direction. If, for example, a body moving freely at constant velocity is acted upon by a force in the *direction of travel*, the result is an increase in velocity. If, however, that force acts at *right angles* to the line of flight, the consequence must be a change in direction. Since the planets are moving around the Sun, a continual change of direction is taking place. This in turn implies a steady force (centripetal) toward the center of the principal focus of the ellipse (i.e., the Sun). In other words, the pull (gravitational attraction) of the Sun is causing all the planets to curve inward toward the Sun. In essence they are continually falling toward it just as the Moon is continually falling toward the Earth. The overall effect in each case is that the bodies continue to orbit their respective primaries.

With this point cleared up, Newton was able to turn his attention to Kepler's three laws of planetary motion. By integrating these with his then recently deduced formula establishing the effect of a central force on an orbiting body, he decided that the force he sought obeyed the inverse square law, i.e., the strength of the force varies inversely with the square of the distance. Thus the hitherto unidentified force was one directed toward the center of a massive body. This, so far as the Solar System was concerned, must be the Sun. Newton came quickly to the conclusion that gravity, that force so familiar on Earth, was in fact the force that had so long been sought. One might say that until that moment science had been unable to see the forest for the trees.

Bear with me a moment while I echo a paragraph from Newton's oft-cited but rarely quoted *Principia Mathematica:*

> The mean distance of the Moon from the Earth is about 60 (Earth) diameters. So if we imagine the Moon, deprived of all its motion to be let go, so as to descend toward the Earth with the impulse of all that force

> by which it is retained in its orbit, it will, in the space of one minute, describe in its fall 15½ feet. Since that force in approaching the Earth increases in the proportion of the inverse square of the distance and, upon that account on the surface of the Earth is 3,600 (60 × 60) times greater than at the Moon, a body in our regions falling with that force ought in the space of one minute of time to describe 60 × 60 × 15½ feet and with this very force we actually find that bodies here upon the Earth really do descend and therefore the force by which the Moon is retained in its orbit is that very same force which we commonly call gravity.

You get the drift, I think, but now you realize *why* some classics of science are rarely quoted.

With gravity finally established as the long-sought force, another fundamental question about the universe remained to be resolved. This related to the *nature* of space. Was it composed of the mystical œther or was it simply a vacuum? The then-prevailing belief about a vacuum was that it was "something nature abhorred," to use the popular expression. To philosophers such as Descartes the medium was the œther, but just precisely what that diffuse substance was seems never to have been adequately explained; the term denoted a medium which philosophers of the period felt *ought* to be there. Kepler, as we have seen in his reply to Galileo, had indicated his belief that space must be a vacuum.

Be that as it may, Descartes and his friends simply could not conceive of the "existence" of nothingness. To them it seemed an insult to intelligence and contrary to all rational thought. Space must surely be full of some all-pervading medium and that something was the œther. Newton rejected this view with vehemence in his *Principia Mathematica:* "I have no regard in this place to a medium, if any such there is, that freely pervades the interstices between the parts of bodies." Neither does it appear that Newton had any doubts about the nature of time. He wrote: "Absolute, true and mathematical time, of itself and from its own nature, flows equally without relation to anything external and by another name is called duration." So here we have what seemed to be the definitive pronouncements of the father of classical physics on the "elements" of the space-time continuum.

But, not surprisingly, the idea of a pervasive œther did not die there and then. Long-held beliefs rarely disappear overnight. In fact, the notion endured for some time and even Newton, as he grew older, appears to have modified his earlier rejection of the œther. Moreover, although he had established that gravity was the force long-sought, his concept fell somewhat short of present-day thinking. In a later book he denied that the cause of gravity lay in the mass of an object and claimed instead that that force was a property of space itself or, as he put it, "of that great sea of œther which fills the universe." Newton's contributions were vital and real enough, so it would be wrong to quibble over an odd bit or two of conceptual and semantic inconsistency.

It may be that Newton had become influenced by a new debate that had by then arisen. The nature of gravity had been solved, the nature of light had not. Light is another fundamental facet of the universe, and in a later chapter, when we deal with possible time travel into the past, we will see how oddly light behaves under the tremendous gravitational pull of a black hole.

In Newton's era virtually every theory regarding the transmission of light invoked the presence of some indefinable elastic "fluid" enabling light rays to traverse space. Light, it was then believed, could *not* pass through a vacuum.

In 1666 Danish astronomer Ole Roemer at the Paris Observatory, using the four principal moons of Jupiter, succeeded in determining the velocity of light. After a few abortive attempts he arrived at the value accepted today: slightly in excess of 186,000 miles per second (300,000 kilometers per second). The velocity of light was therefore *not* infinite as had been widely believed.

The investigation of light now centered on two important questions: (1) In what manner was light transmitted through space? and (2) Exactly what was light? Besides that of Newton, the name of Christiaan Huygens (1629-1695) stands out in this early scientific quest. The theories advanced by the two men were diametrically opposed. We'll not go into great detail regarding these theories since that would digress too far from our basic theme. Suffice it to say that Newton regarded light as a stream of very fast-moving particles, giving rise to what was called his Corpuscular Theory. Huygens, on the other hand, produced his Wave Theory, which was just as plausible. His response to Newton's stream of particles was that, since light rays could cross through each other, surely if they were composed of small material particles, innumerable collisions must occur. Waves, however, were an entirely different proposition, and Huygens likened light waves to waves in water. Fling a pebble into a pool of water. The waves pass concentrically outward, but in fact the water itself is *not* in outward motion. To put it another way, energy is passing through a medium, which in this case is water, but it must be emphasized that the medium itself is *not* moving laterally. This analogy with water may be unfortunate inasmuch as a medium of transmission is being invoked, i.e., water. Whereas Newton believed light to be composed of solid particles, Huygens maintained that the œther was composed of myriads of exceedingly tiny particles in which the waves had the effect of a percussion flow passing through them at high velocity, the effect being similar to that produced when a stream of marbles hits a line of stationary marbles.

For more than a century these theories remained irreconcilable. Newton again abandoned the concept of the œther. To him space once more was a vast vacuum in which gravity was capable of manifesting its effects over immense distances and through which passed fast-moving particles of light. Huygens, no doubt under the strong influence of the Descartes school, envisaged space as a great ocean of œther through which light could spread in a wave-like fashion due to the "knock-on" effect on small particles of the supposed œther. Whichever theory came to be accepted might well answer the mystery of the nature of space.

In the event both Newton and Huygens proved wrong, and an entirely new wave theory was eventually propounded by Dr. Thomas Young, professor of physics at the Royal Institution in London. Young postulated that light had *not* the type of wave form in which a compression travels through a medium, i.e., a longitudinal wave. It was, he maintained, a *transverse* wave in which a lateral or sideways ripple spreads through the intervening medium. Now transverse waves are only able to traverse a medium in *tension,* i.e., some form of elastic solid. Their existence in liquids or gases is impossible since both are incapable of springing back into their original shape. Taken to its logical conclusion this implied that the medium in which the stars and other celestial bodies moved with such consummate ease must be a *solid.* This was plainly ridiculous. Space, then, simply had to be a vacuum. The œther, that long-enduring medium so dearly loved by Descartes and others, was merely an abstraction; it could not be real.

Where has science brought us in the many years since this pioneering work? Where have we come in our study of what we will term, for want of a better, the conventional universe? It is one populated by vast galaxies in the great immensity, and these contain not only stars but great clouds of dust and elemental hydrogen. Many of the stars are likely to have planetary systems not greatly unlike that of the Sun. All are surrounded by an almost total vacuum, and their movements are dictated by gravity. The light from them is transmitted by wave action. Here then, in a cosmic nutshell, is the universe as most of us see or understand it. But having apparently come to terms with it, it is not altogether surprising if other questions begin to arise. Is this immensity finite? We have finite minds that exist in a finite world, so it is understandable if we tend to regard the universe also as finite. And this is precisely where the real problems commence. If the universe is finite, it must somewhere have a boundary. But if so, what lies beyond that boundary? "Nothing," we may be inclined to reply, but how does one visualize or define *that* sort of "nothing"? Surely the black intergalactic void of space is nothingness. So what lies beyond the boundary of space, assuming there is one? What kind of nothingness is that, and does *it* go on forever? Perhaps something does lie beyond this boundary. But if so, what, and does it in turn have a boundary beyond which presumably lies something else? The problem speaks for itself. Let us, on the other hand, assume the universe is infinite, that it just goes on forever and ever—an endless succession of great galaxies frightening in their remoteness. At the edge of that alien and existential abyss, most of us simply shrug in recognition of our incapacity.

When we begin to broach these difficult questions, we come to an entirely new and different cosmology—the universe not of Newton and his contemporaries but that of Albert Einstein (1879-1955). In this strange and almost unreal universe, gravity takes on another dimension entirely, one that leads to black holes and probably to time warps. These and other closely related aspects are the basis for the next two chapters. In them we will see how space and time are inextricably linked.

RELATIVITY

Before we examine the much stranger universe of Einstein and its bizarre implications, we should be at least acquainted with the concept of relativity.

The first point that must be clarified—and there seems to be some confusion about this—is that Einstein produced *two* theories of relativity—his *special* theory of 1905 and his *general* theory, propounded in 1916. The two cannot be considered synonymous. We will therefore deal with them in chronological order.

Special Theory of Relativity

The Special Theory of Relativity involves the laws of physics as they apply to observers moving relative to one another at a *constant* velocity. They are generally referred to in the literature as observers in *inertial frames.* An inertial frame of reference is simply one in which an isolated body at rest will remain so, whereas a body in uniform motion will maintain a constant velocity. It is therefore a frame of reference *not* subject to acceleration as, for example, from a gravitational field. Let us now see how the relative motion of observers affects the measurements they make.

The Special Theory predicts that if velocities are of a low order the results will be in accordance with the conventional Newtonian laws of mechanics. However, as relative velocities increase far beyond those we can possibly regard as normal (i.e., speeds close to that of light), the laws of Newton and Einstein diverge dramatically. Although one might be forgiven for relying on what we call "common sense" and regarding Einstein's conclusions with a measure of skepticism, it is fair to state that they have been verified fairly conclusively by experiment. It is therefore quite in order to use special relativity when dealing with particles of an atomic nature—e.g., electrons—that move at velocities close to that of light. The velocity of light is a constant denoted by the letter "c." In a vacuum it is precisely 299,792,458 meters per second. Space, however, is not quite a total vacuum, and this tends to lower the velocity very slightly. It is reduced slightly more when light enters the Earth's atmosphere. For most practical purposes, though, the velocity of light can be expressed as Roemer calculated it: 300,000 kilometers per second, or 186,000 miles per second. Now one of the fundamental tenets of the Special Theory is that the velocity of light in free space is the same in *any direction* as well as being the same for each and every *observer.* It is also quite independent of the *motion* of the body emitting the light. Although this appears to violate the accepted concepts of relative motion, the facts have been proven experimentally.

Prior to the advent of Einstein's Special Theory the belief was that absolute time and absolute space were identical to *all* observers. Einstein changed all this by demonstrating that, as a direct consequence of the constancy of light velocity, *space and time could no longer be regarded as disparate and independent entities.* They were, in fact, constituents of a composite entity termed *space-time,* a term we will use fairly frequently in the pages to follow. Space-time is simply a single entity in which space and time are *unified.* Thus the location of a body in space can be specified by four instead of three coordinates. Three of these give its position in space (an entirely conventional and understandable proposition) and one its position or point in *time.* The course of the body in space-time is generally referred to as its "world line," and this links events in its history. When we proceed to the General Theory of Relativity we will see how the presence of gravitational fields produced by stars and other bodies causes space-time to become *curved.* So far as the Special Theory of Relativity is concerned, however, space-time may be regarded as *flat.*

Now for the first time we come across something that could yield the key to time travel and allow us access to the future. This would involve an interstellar round-trip at a velocity close to that of light; we'll examine that at length in a later chapter. As the velocity of the spacecraft, v tends toward the velocity of light, c, a very odd situation begins to develop—the phenomenon known as *time dilation.* Put briefly, a pair of observers approaching each other at a velocity close to that of light would, were it possible for them physically to do so, each see the chronometer of the other advancing more *slowly* than their own. In fact the time intervals would be lengthened (dilated) by a factor of $1/x$ where x is equal to

$$\sqrt{1-\frac{v^2}{c^2}}$$

The result of applying this important formula will be explained later. For the moment suffice it to say that its terms might open the way to the stars as well as into the future.

The Special Theory produces another bizarre consequence. The mass, m, of a body in motion is *not* invariable. It *increases* as the relative velocity between a body and an observer increases. The relevant formula is as follows:

$$m = \frac{m_r}{x}$$

where m is the mass of the body in motion, m_r the mass of the body at rest, x as before is equal to the expression

$$\sqrt{1-\frac{v^2}{c^2}}$$

It must be stressed, though, that this increase in mass becomes apparent only at velocities close to that of light; for example,

$$m = \frac{m_r}{x} = \frac{m_r}{\sqrt{1-\frac{v^2}{c^2}}}$$

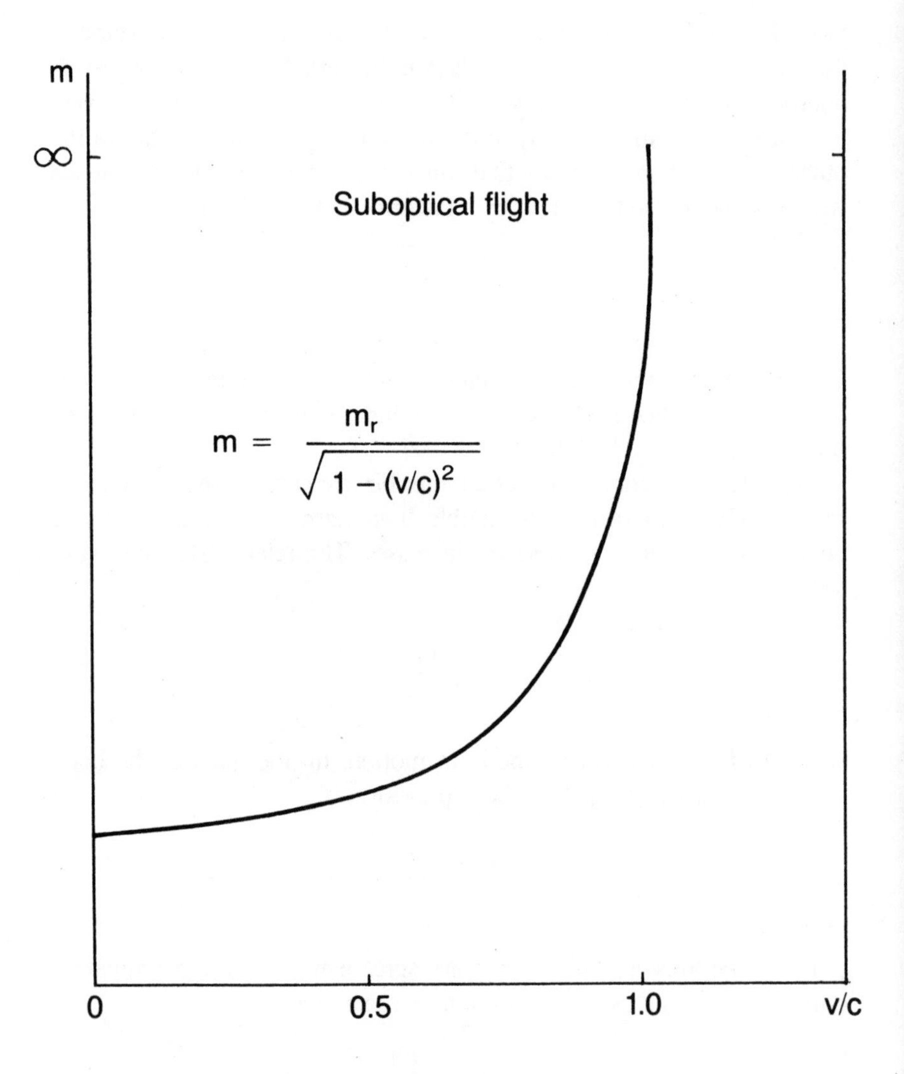

Figure 2a
Graphical representation of the rate at which mass increases with velocity.

As v tends toward c, then

$$\sqrt{1 - \frac{v^2}{c^2}}$$

has a very low value, (i.e. x has a very low value) and therefore

$$\frac{m_r}{x}$$

has a high value. And that means that m has increased due to the high value of v.

Conversely, if v is very small compared to c, then

$$\sqrt{1 - \frac{v^2}{c^2}}$$

is very nearly equal to 1, and so, accordingly, is x very nearly equal to 1. Thus $m \simeq m_r$ (little change), which is what we should expect.

One notable consequence of this is that *no object with mass* can attain the velocity of light. The mathematics are fairly straightforward:

$$m = \frac{m_r}{x} = \frac{m_r}{\sqrt{1 - \frac{v^2}{c^2}}}$$

But if $v = c$, then $v^2 = c^2$ and

$$\sqrt{1 - \frac{v^2}{c^2}} = 0$$

i.e.,

$$m = \frac{m_r}{0} = \infty$$

Thus, m has *infinite mass* at light velocity, which is clearly an absurdity.

However, a particle with zero rest mass ($m_r = 0$)—e.g., a photon—*can* travel at the speed of light (i.e., $v = c$).

$$m = \frac{m_r}{x} = \frac{0}{x} = \frac{0}{\sqrt{1 - \frac{v^2}{c^2}}} = 0.$$

That is, at light velocity, m (mass in motion) is zero.

The ability of photons to travel at the velocity of light is hardly surprising, since a beam of light is in fact a beam of photons.

Rest mass is simply a form of energy capable of being converted into alternative forms of energy. The relationship between mass and energy is expressed by the well-known equation $E = mc^2$, E being the total energy of the particle. The velocity of light is so great that the equation implies a vast release of energy for a minute loss of mass—approximately 30 million kilowatt-hours of power per gram of matter. Any doubts concerning this were removed in 1945 when the first nuclear device was detonated.

General Theory of Relativity

In a sense the General Theory of Relativity is a particular extension of the Special Theory, for it demonstrates how the relationship between space and time is affected by the gravitational effects of matter. Gravitational fields alter the geometry of space-time, causing it to become *curved.* As will be shown later, the rules of geometry as they apply to curved space cannot be equated with those of normal Euclidean geometry. In the Special Theory the parameter of gravity was ignored and therefore space was "flat." The curvature of space-time effectively controls the motions of celestial bodies. In fact it could be said that matter "instructs" space-time how to curve, and in its turn space-time "instructs" matter how to move. In the preceding chapter we dealt at some length with the Newtonian universe, in which gravity exerted a profound and fundamental influence. Readers may be tempted to wonder how the Einsteinian universe differs since it too is partly based on a theory of gravitation. The essential difference is that Einstein postulated the effects of gravitation "in extremis," intense gravitational fields in the vicinity of stars and black holes.

Einstein showed that the natural motion of a body is a straight line. In other words it follows the shortest distance between two points. This is known as a *geodesic.* So far all this sounds very much like basic high school geometry. However, there are complications that render the conventional geometry of Euclid invalid. A simple diagram illustrating this point (Figure 6) appears in a subsequent chapter. What we are confronted with is the geometry of space-time rendered curved by reason of strong gravitational fields. In fact, the actual degree of curvature produced by a particular gravitational field can be calculated from Einstein's field equations. It thus becomes possible to calculate the geodesics followed by bodies moving in curved space-time.

Experimental confirmation of the General Theory of Relativity involves either (a) extremely accurate measurements, or (b) the existence of a very strong gravitational field. Experiments must substantiate the predictions of this theory in instances where there is a deviation from those of the more conventional (Newtonian) gravitation. So let's look at three experiments that have been performed and at their results.

1. Light rays (and all other forms of electromagnetic radiation, e.g., radio waves) are bent in a gravitational field. During a total eclipse of the Sun, for example, the apparent positions of background stars then lying close to the Sun's limb differ slightly in their ascension and declination from the positions they normally occupy when they lie far from the Sun's limb. This is entirely due to the gravitational effect of the Sun's mass on the light rays from the stars. The difference is 1.74 arc seconds, a figure

within 5% of the deflection predicted by Einstein. The bending effect also is manifested on radio waves (which differ from those of visible light only in their much lower frequency). In fact, still closer agreement with Einstein's predictions has been found with respect to microradio waves emanating from radio sources deep in space. More recently observers have recorded that there is a time delay in radio signals transmitted by spacecraft when they are close to the Sun. These delays have been measured accurately and substantiate Einstein's predictions within 3%.

2. When electromagnetic radiation (and this of course includes light) is emitted from a body as massive as a star, there occurs a shift in the lines of the spectrum. This is known as *gravitational red shift* (or sometimes as Einstein shift). The reason for this effect is that in escaping from the body's strong gravitational field the radiation loses energy. As a direct consequence the frequency of the radiation decreases. This means that the wavelength increases. (In the visible spectrum, red has a lower frequency and longer wavelength than violet, at the other end.) For those interested in the relevant mathematics, the extent of red shift is indicated by the following equation:

$$\Delta\lambda = \frac{Gm\lambda}{c^2 r}$$

Where G is the gravitational constant, m and r the mass and radius of the massive body respectively, c as before is the velocity of light, λ is the wavelength, and $\Delta\lambda$ is the amount of wavelength shift.

The predicted amount of gravitational red shift (toward the red because as frequency decreases, wavelength increases) in the spectral lines of radiation being emitted from a massive body has been demonstrated successfully with respect to both the Sun's and the Earth's gravitational fields. Since we live upon the Earth, it is possible to prove relativistic effects in another way. Clocks should run more slowly in a strong gravitational field than in a weak one. This being so, we should expect a difference in the time-keeping properties of clocks at widely varying altitudes. One close to the surface of the Earth would be expected to run more slowly than one at a high altitude, for the nearer to the planet's surface, the stronger the gravitational field. Obviously the discrepancy will be exceedingly slight, and to detect it necessitates the use of extremely accurate and sensitive instruments. For the record, an atomic clock flown at an altitude of 10 kilometers (6.25 miles) ran 47×10^{-9} seconds faster than one flown just above sea level—almost exactly the difference predicted by the equations of Einstein.

3. The effect of curved space-time influences the motions of orbiting bodies such as planets. This has been demonstrated fairly conclusively in the case of the planet Mercury when at perihelion. Because the planets pursue elliptical orbits, there are occasions when they are nearest to the Sun (perihelion) and others when they are farthest from it (aphelion). (As a matter of interest, Earth reaches aphelion around the 3rd of July each year—close to midsummer in the northern hemisphere.) The perihelion point of a planet's elliptical orbit shows a gradual movement in the same direction as that of the planet's orbital motion. This is due in part to the gravitational influence of other planets, but also to the *curvature of space-time in the vicinity of the Sun.* The contribution due to the latter is slight but, as we shall see a little later, was predicted by Einstein.

Einstein's two famous theories fairly revolutionized scientific thinking when they were propounded in the early years of this century, and they have continued to do so ever since.

Relativity, by its very nature, is not one of the easiest subjects to comprehend, far less to master. The brief recounting of those two companion theories in this chapter does no more than skim the surface. Nevertheless, any serious discussion of time-travel would be impossible without frequent reference to them. Though it would require an entire volume to treat relativity in the depth it warrants, the fundamentals expressed here should prove helpful when we come to examine time dilation, near-light velocities, and the possibilities of faster-than-light travel, for these things, by their nature, are fundamental to time travel. It is a peculiar, unconventional universe at which we will be looking, and if we're to see clearly we mustn't permit our minds to be fettered or clouded by traditional motions of common sense and rationality. Time, like length, breadth and depth, is a dimension that can be changed—but with much less ease.

THE "NEW" UNIVERSE

Newton, we already know, regarded gravity as some kind of force that emanated from the Sun and the planets. In a sense this is correct. If we let go of a solid object here on Earth, it falls to the ground. Einstein, however, saw things rather differently. He was convinced that gravity was not so much a force, but a *distortion of space-time,* with effects hitherto undreamed of. Rays of light, as we have seen, will not travel in straight lines in regions of strong gravitational attraction. And time will run *more slowly* in certain circumstances. These are attributes of distorted regions of space-time and of bodies traveling at near-light velocities. Repeating once more, space is not just space—it is indeed space-time, a continuum. And by way of good measure, we have to accept the fact that space, due to strong gravitational fields, is curved, especially in the vicinity of massive celestial bodies. But how, you may justifiably ask, can space, which is "nothingness," be endowed with a property rightly belonging to something tangible? That's the problem we'll address first.

A basic axiom of Euclidean geometry is that the shortest distance between two points is a straight line. On a flat sheet of paper or the like this is an indisputable fact, easily proven. It also seems to be true between two points on the Earth's surface. But is it? Remember that the Earth's surface is curved. If the two points are very close together, the maxim still largely applies. But suppose that the two points are represented by two cities 100 miles apart. What then? The curve between them may be short and of tremendous radius, but it is a curve nevertheless.

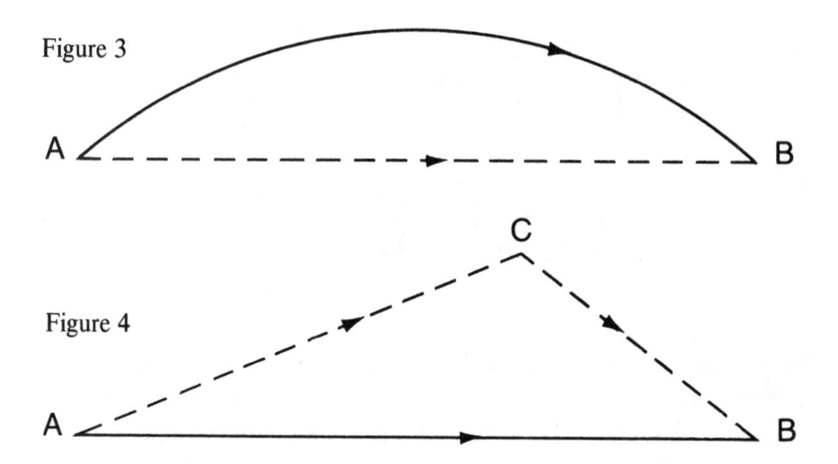

A glance at Figure 3 shows that although the shortest distance between cities A and B, 100 miles apart, may have all the appearance of a geometrically short path, that cannot really be. Viewed on a map (plan elevation) it assuredly looks that way (Figure 4) and certainly proceeding from city A to city B via city C would involve a marked increase in the distance traversed. But as Figure 3 shows, the *real* straight line and thus the *shortest* distance between A and B must be a "subterranean" route. Perhaps if such a tunnel were cut and the design engineers paid due regard to the configuration imposed by the radius of the Earth, the *real* straight-line path could be followed (though at exorbitant trouble and expense). When we come to consider long distances over the Earth's surface, say, for example, from New York to San Francisco (3,000 miles), the truth of this becomes even more evident. Indeed if we take the thing to its logical (and ridiculous) conclusion and envisage a journey from a point on Earth's surface to its exact opposite (the antipodes), the shortest distance geometrically is right through the center or core of the Earth. It would seem then that the standard geometry of Euclid collapses.

During the early years of the nineteenth century a noted German mathematician, K. F. Gauss, produced an entirely new system of geometry in which lines were drawn not on flat surfaces but on curved ones. The results were highly interesting and instructive. Diverging lines now *met* and squares no longer had angles of 90 degrees. These phenomena contradicted those of Euclid, who had restricted himself to flat surfaces.

Figures 5 and 6 provide practical examples. In Figure 5 a triangle is drawn on a perfectly flat plane. No matter how far lines CA and CB are extended, they cannot possibly meet. On the contrary they will continue to diverge. If, however, as in Figure 6, the triangle is drawn on a spherical surface, they must eventually meet at D.

Although Gauss was tempted to treat these facts more as an amusing mathematical paradox than anything else, his one-time pupil Bernard Riemann developed them into a system of curved geometry embodying

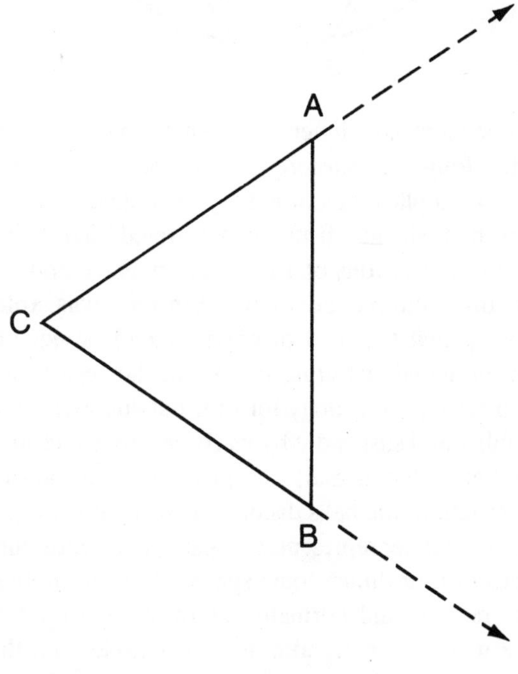

Figure 5

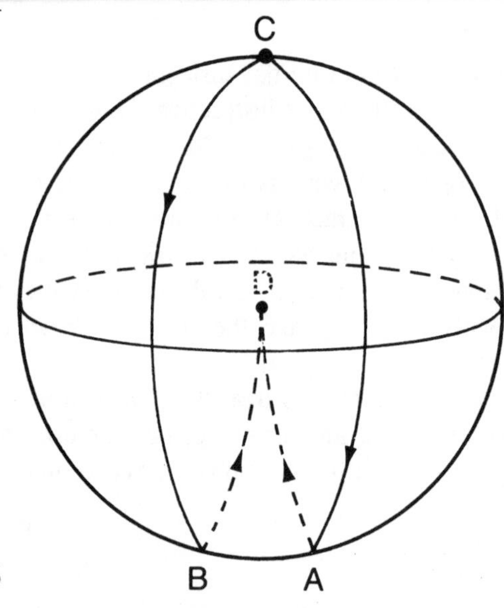

Figure 6

not only the three "normal" dimensions (length, breadth, and depth) but also the essential fourth dimension, time. It was this that Einstein used to explain the true implications of gravity, including especially its effect on light, for it had already been demonstrated that light rays were somehow *bent* in the environs of a large and massive body in space. By utilizing Riemann's curved geometry, Einstein was able to derive equations showing that the field of gravity could be equated with the precise amount of actual curvature in a particular region of space-time. It is not easy to devise an analogy for this, but suppose we take a large sheet of thin rubber and suspend it by its corners over a clear space. Then we take a number of tennis balls and place them at various points on the sheet. The weight of the balls distorts the sheet at these points. In our analogy the sheet of rubber represents the space-time continuum, the tennis balls, stars. Thus in four-dimensional space objects in motion, star-ships as well as light rays, would normally follow straight lines, but if space is curved, these would be more akin to "great circles" on the surface of the Earth. A star-ship far from a star or any other massive celestial body would follow a path that seems to us perfectly straight. If the star-ship then approaches the environs of a massive celestial object, its "straight-line" path will distort into a curve because the space-time dimension itself, in which the ship is traveling, is distorted by the huge "near-by" mass.

This concept is clearly at variance with that of Newton, who supposed that individual celestial bodies exercised a *general* influence throughout space. Einstein's belief was that such bodies exercise a *localized, regional distorting effect.* To illustrate this more clearly let us return to the sheet of taut rubber in which several tennis balls have settled in depressions due to their weight. Now suppose we take a ping-pong ball, which is very light, and roll it at random over the rubber sheet. If it does not closely approach a "tennis ball depression," it will continue on its way unperturbed; but if it does enter the vicinity of a "tennis ball," then the ping-pong ball will accelerate, change direction, and spiral into the depression. Now let's recall the planets orbiting the Sun and look at an interesting feature not entirely unrelated to our model. The ping-pong ball will eventually come to rest against the tennis ball; the Sun, just another star, approximates that tennis ball. In the Sun's vicinity gravity is very strong, but the planets, fortunately for us, do not fall into that star but orbit around it.

It is obvious that the region of gravity-induced curvature is greatest close to the body responsible for it. In regard to the Solar System, then, we would expect the planet closest to the Sun, Mercury, to show the effects most strongly since it is a mere 33 million miles from the Sun. Newton's laws predicted that planetary orbits should be elliptical, and this was eventually found to be so. It was discovered more than a century ago, however, that Mercury was *not* conforming exactly as expected. The discrepancy was very slight, but it was real. This aforementioned feature is known as the *advance of the perihelion.* This is the gradual movement of the perihelion of Mercury's elliptical orbit in the same direction as that of the planet's orbital motion. It is due not only to the gravitational influence of other planets but also to the curvature of space-time around the Sun as was predicted by Einstein's General Theory of Relativity. If the path could be traced in space, it would appear roughly as shown in Figure 7, although in the interests of clarity the divergences have been greatly exaggerated. Despite the slightness of the discrepancy (about 43 arc seconds per century) its existence must have encouraged Einstein greatly since his theory had predicted an effect of this order.

Einstein's General Theory of Relativity had a very profound effect upon the scientific world, upsetting as it did so many long-established beliefs. Celestial bodies such as great stars had to some extent been shorn of their glory. Now it seemed they were merely masses of matter responsible for creating local distortions in the fabric of the space-time continuum. Gravity is to all of us a very real thing. If we fall off a cliff, we immediately drop downward. This of course is due to Earth's gravity. But the Earth is a relatively small body by celestial standards, and it is not altogether easy

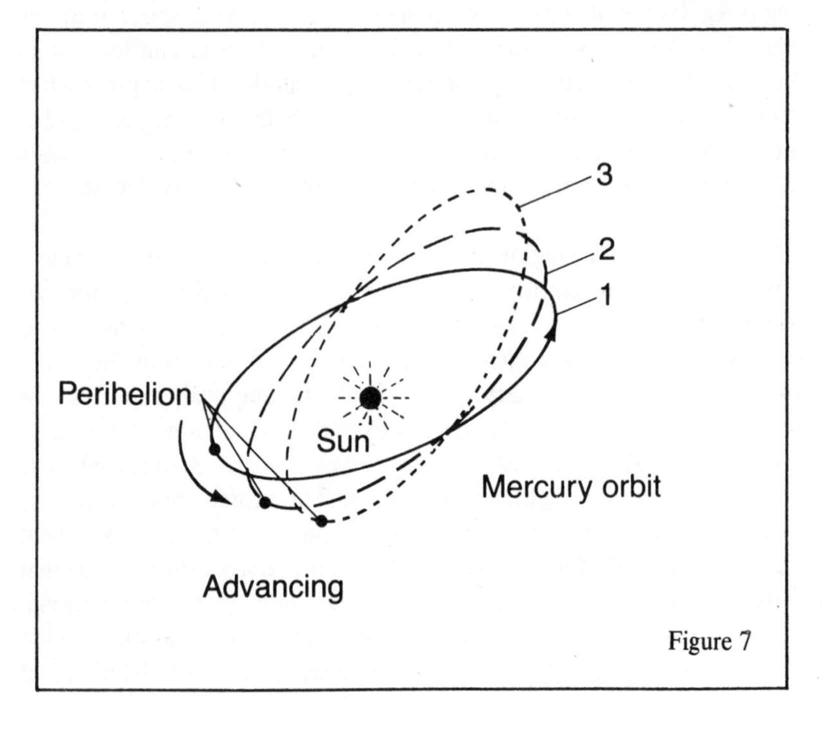

Figure 7

to attribute such a fall to the Earth's having caused a curvature in space-time.

The more we examine the universe of Einstein the more we realize what a strange, unusual place it is. Eventually, we will come to the subject of black holes where the effects of gravity are so intense that they trap and retain the light rays emanating from massive collapsed stars. Black holes are of such special interest with respect to time travel that they will be dealt with at length in a subsequent chapter.

At this point we should briefly mention *gravitational waves.* These are thought to be due to remote stars that experience a sudden change in shape for one reason or another. Such waves had been predicted by Einstein, who regarded them as *changes in the curvature of space-time* likely to be propagated through the universe at the velocity of light. Just as electric and magnetic forces have a wave aspect known as electromagnetic radiation, so gravitational force should presumably have its wave aspect, i.e., gravitational radiation. Although predicted by the General Theory of Relativity, this line of speculation has yet to be confirmed experimentally. Gravitational waves, it is thought, are most likely to originate where large masses are being accelerated rapidly. The most likely sources would be supernovas, embryo black holes, and binary pulsars. Whereas electromagnetic waves affect only charged particles, gravitational waves should affect all matter. Many physicists are today strongly convinced that such radiation does exist.

Einstein eventually applied his beliefs regarding the curvature of space-time to embrace the entire universe. Earlier it had seemed that curvature would not exist in those regions of space far removed from the powerful gravitational attraction of stars, and that therefore any small interstellar particles could travel endlessly in straight paths. Our model, with its taut rubber sheet, tennis balls, and ping-pong balls gave this impression. We will see now that this model, though it served a useful purpose, was *not* really accurate in the full sense. If matter and light were to proceed endlessly in straight lines forever, the universe would surely run down because of the losses involved. After all, our hypothetical rubber sheet had boundaries, and the ping-pong balls that escaped the tennis-ball depressions must eventually go over the edge and be gone irretrievably. If, however, it were possible to have a *spherical* rubber sheet with embedded tennis balls, the ping-pong balls that missed the depressions would eventually reappear at their starting point. (Think of a fly crawling around a suspended sphere.) That sort of four-dimensional space-time allows us to visualize the same thing with respect to matter, beams of light, and other forms of energy. Such a universe would *not* run down because there would be no losses. The curvature involved would be slight, but Einstein's equations prove that it does curve, more so in those parts of the universe where the density of matter is greatest and vice versa. It follows that if the mean density of the universe is high enough, space-time must be bent around on itself; space-time must be a closed system.

Some astrophysicists, however, saw a curved closed universe as inherently unstable, the mathematics of the situation leading them to believe that it would either contract to virtual nothingness or expand indefinitely. Further research would show that there was at least something to one of these notions.

Among the stars, if slight optical aid is employed, can be found at various points faint, fuzzy patches of light that are clearly not stars. With much greater optical aid some of these are seen for what they are—great glowing clouds of gas. A classic example is M42 (its number in Messier's famous catalogue), better known as the Great Nebula in Orion. This vast cloud of elemental gas and dust contains several "embryo" stars, and these illuminate the great cloud of gas. However, other nebulas were found to show an entirely different form, being in the main spiral configurations resembling gigantic "Catherine Wheels" and suggestive of slow, majestic motion. Now, nebulas such as that in Orion lay indisputably within our own galaxy (the Milky Way), and for a time there seemed no reason to doubt that the peculiar spiral objects also lay within the confines of the Milky Way. At this point astronomy was poised to take one of the greatest leaps forward in its history. The eminent philosopher Immanuel Kant (1724-1804) had proposed that, far from being an integral part of the

Milky Way, these spiral (and sometimes elliptical) objects lay far *outside* it and were in fact great galaxies in their own right or "island universes." Kant's view was not then widely accepted, and most astronomers regarded it as the sheerest speculation since there was no valid evidence to justify it. One fact, however, continued to stand out in a puzzling way. While the distances from the Earth to the stars could be measured (and this included that of objects like the Great Nebula in Orion), it proved impossible to measure the distances to these other strange objects. Moreover, spectroscopic examination of the gaseous nebulas showed the typical effect to be expected from a very hot cloud of gas. When the other objects were subjected to this form of examination, the results were markedly different. Their emitted spectra evidenced a number of fine, dark lines similar to those shown by the Sun and other familiar stars, these dark "absorption lines" due to the presence of specific chemical elements in the star's atmosphere. This was certainly odd, and the only conceivably valid construction to fit the facts was that the spiral galaxies were systems of stars in their own right; our inability to measure the distances to them was simply because they lay well outside our own galaxy, so remote that the technology of the period was unable to cope. Kant's hypothesis had been vindicated.

Consider the truly immense implications. At a stroke the confines of the known universe had been extended almost beyond comprehension. The Milky Way was just one galaxy, the one to which our Sun and 100,000 million other stars belonged. Beyond it, at distances from which the mind reeled, lay other immense galaxies of stars, some much greater than our own. If this were not enough, another tremendous discovery was soon to follow. In 1917, Edwin Hubble (1884-1953), at Mt. Wilson Observatory in California, observed that absorption lines in the spectra of several spiral nebulas then under observation were being displaced toward the red end of the spectrum. This classic example of the renowned Doppler shift could be construed in only one way—the galaxies concerned were moving *outward* and away from us at fantastic velocities estimated at 400 miles per second. Later, in 1929, Hubble discovered that all the external galaxies were receding at velocities directly proportional to their distance from us. The recession of the galaxies had been established—the universe was expanding!

Just what order of distance was involved with respect to these other galaxies? Once again it fell to Edwin Hubble to provide the answer. By the mid-'30s he had devised the necessary techniques. The result was, to put it mildly, staggering—at least 500 million light-years, and even this seemed likely to be a conservative figure. The nearest galaxy turned out to be M31, the Great Galaxy in Andromeda, a mere 2.25 million light-years from us. Today we know the calculation of 500 million light-years was grossly conservative. One spiral galaxy in the constellation Hydra, for example, is 2,600 million light-years distant, and it is by no means the most remote. Let's let Edwin Hubble speak for himself in this brief passage from his fascinating 1936 book, *The Realm of the Nebulae:*

> The Earth we inhabit is a member of the solar system—a minor satellite of the Sun. The Sun is a star among the many million which form the stellar system. The stellar system is a swarm of stars isolated in space. It drifts through the universe as a swarm of bees drifts through the summer air. From our position somewhere within the system we look out through the swarm of stars past the borders into the universe beyond.
>
> The universe is empty for the most part but here and there, separated by immense intervals, we find other stellar systems comparable with our own. They are so remote that, except in the nearest systems, we do not see the individual stars of which they are composed. These galaxies are scattered at average intervals of the order of two million light years or perhaps two hundred times their mean diameters. The pattern might be represented by tennis balls fifty feet apart. The order of the mean density of matter in space can also be roughly estimated. It is about one grain of sand per volume of space equal to the size of the Earth.

Just how fast are the galaxies receding? Early estimates were of the order of a few hundred miles per second. Later, when the spectra of more remote galaxies were examined, the velocities proved to be in excess of 1,000 miles per second or 3.6 million miles per hour. It was as if our own galaxy had developed some sort of cosmic plague and the others were distancing themselves from us as fast as physically possible!

Hubble was now able to coordinate the figures for galactic distances with the appropriate velocities. His results provided a simple mathematical law valid throughout the entire universe. Put briefly, it was this: *The farther a galaxy is from us, the greater is its velocity of recession.* Indeed, *Hubble's Constant,* as this law came to be known, states that for every additional million light-years from us velocities increase by 20 miles per second. The most remote galaxy yet to have its velocity measured is receding at over 70,000 miles per second, i.e., at about 38 percent of the velocity of light. As if that were not astounding enough, in 1963, Maarten Schmidt at the Mount Palomer Observatory detected a quasar, a "quasi-stellar object," hurtling outward at 80 percent of light velocity—at nearly 149,000 miles per second. This raises a very intriguing question. If there are galaxies even more remote (and the evidence is that there almost certainly are), the velocities of these receding objects could in the end approach that of light itself. What notions must we derive then, recalling that the appropriate equations indicate that the velocity of light can neither be equaled nor exceeded? Do such galaxies or quasars represent the physical confines of the universe?

To all appearances the galaxies are retreating from us, but this is *not* the real state of affairs. Without doubt the distance between the Milky Way and every other galaxy is increasing, but in fact *every* galaxy is distancing itself from *every other* (so after all the Milky Way is not a cosmic pariah). Suppose we were to procure a child's balloon the surface of which was covered with black dots. If we inflate it, we will find that as it grows larger every dot becomes more widely separated from the others. Think of the dots as galaxies and the balloon as the universe, and you'll see how all galaxies separate in an expanding universe (Figure 8).

The question then arose as to whether an expanding universe could continue to expand indefinitely. This seemed improbable. Moreover, there

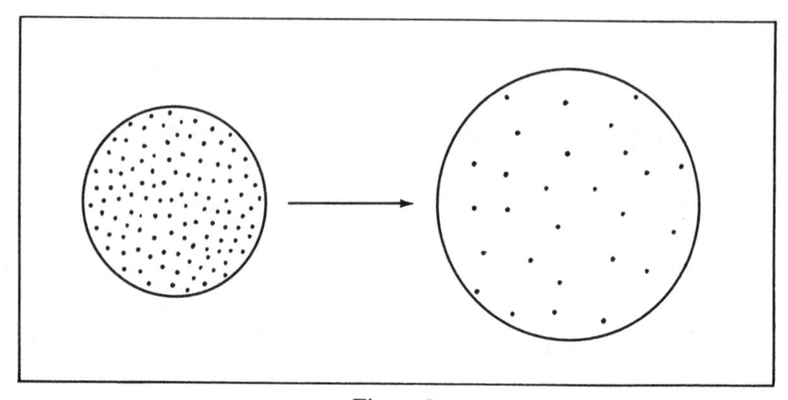

Figure 8

was a second important question: What had started it all in the first place? Since the galaxies are receding, they must at one time have been much closer to one another. The latter question was answered in 1927 by Abbé Lemaître (1896-1966), who envisaged in the beginning a monstrous primeval "atom" the diameter of which may have been of the order of 100 million miles. The matter within it consisted of neutrons, protons, and electrons all tightly jammed together. At some date, it may have been around 20,000 million years ago, this gigantic "atom" (for want of a better term) exploded, and at this precise moment what we understand as space-time began. The contents of this colossal "cosmic bomb" were hurled outward with a violence and fury beyond the power of words to describe. A few minutes later the expanding sphere had reached a temperature of several billion degrees. An hour later, still expanding, this had dropped to 100 million degrees with particles already combining to form nuclei. Thirty million years later the temperature had dropped to a mere few thousand degrees with a certain amount of the gas already condensing into dust. The effects of gravity led to the chance accumulations that in time became the galaxies.

A universe in the process of rapid expansion must be, as we have just seen, very much denser and hotter. This being so, it is reasonable to assume that a considerable amount of "residual" radiation will be emitted as the expansive cooling process continues. According to Jim Peebles and Bob Dicke of Princeton, such residual radiation should be detectable in the form of low-energy radio emanation. In 1965, research workers at the Bell Telephone Laboratories in New Jersey detected a hitherto unknown form of radiation that subsequently proved to be emanating from all quarters of the universe. This may well prove to be the low-energy radiation postulated by Peebles and Dicke.

A primeval cosmic explosion raises another query: How did it originate? This ties in with the question raised earlier: Can an expanding universe continue to expand indefinitely? Today cosmologists are tending to the belief that the "big bang" was not just a "once-only" convulsive outburst but *one of a continuing series.* According to Allan Sandage of Mt. Wilson, these may take place once every 80,000 million years. Eventually the hurtling, outward flight of the galaxies will begin to slow down, grind through aeons to a halt, and then begin to start closing in on one another. The present expanding universe may in time become the *contracting* universe. Ultimately all this dispersed mass of matter and energy could coalesce once more into a renewed cosmic "bomb" that in due course would detonate yet again, these cycles going on ad infinitum.

It is now time to consider the actual *shape* of the universe. So far we have thought more or less in terms of a sphere, but this need not be so and possibly isn't. The solution may involve a possible *fifth* dimension, one that usually is termed "nonspace" or "hyperspace." Other dimensions such as this one (and there may be more!) are of special importance to the theme of time travel.

Curved space, the four-dimensional space-time continuum, is, as we said earlier, difficult to visualize. If it is any comfort, even the most eminent scientists are fully conscious of the difficulty. In his *Mystery of the Expanding Universe,* William Bonnor says, "I do not ask you to visualize ... curved space. I ask you you simply to admit it is a possibility." Likewise in his *Feynman Lectures on Physics,* Richard P. Feynman says, "We live in three dimensional space and we are going to consider the idea that

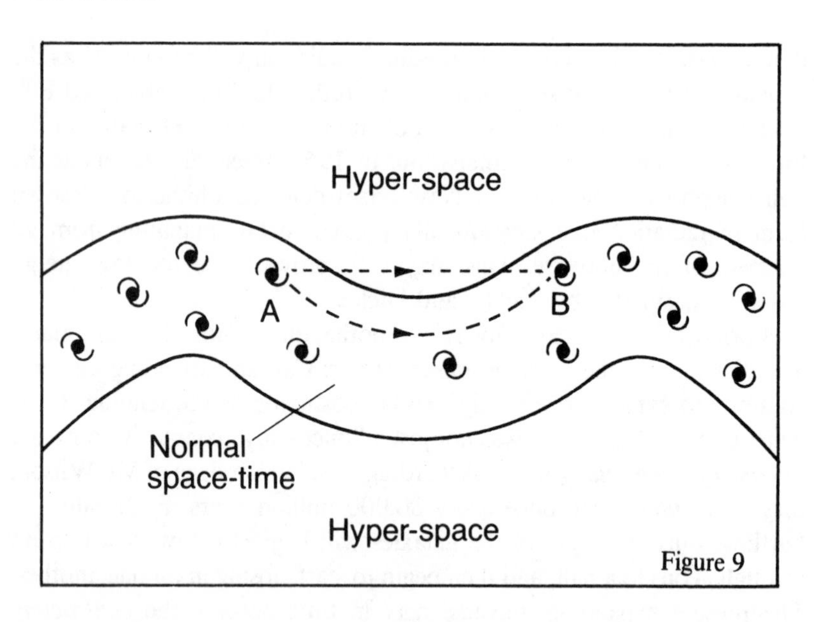

Figure 9

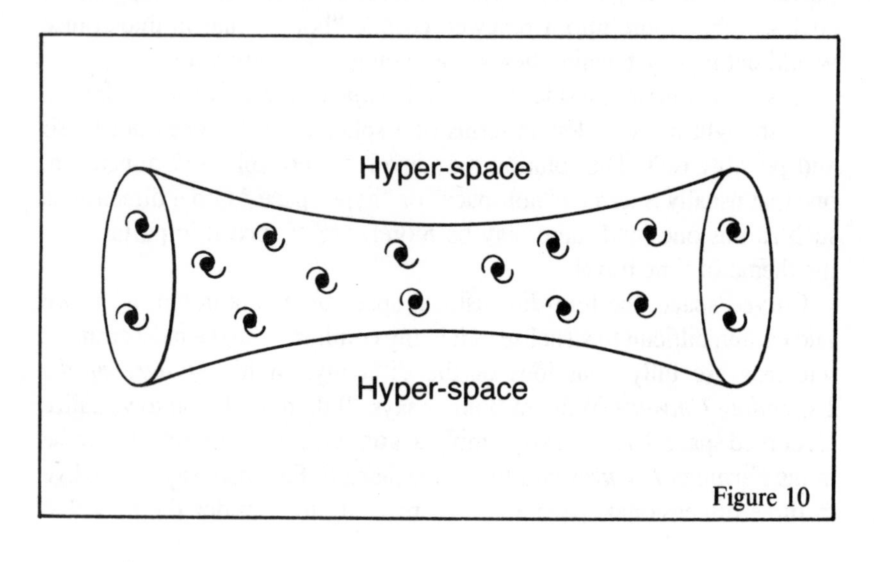

Figure 10

three dimensional space is curved. You say 'But how can you imagine it being bent in any direction?' Well, we can't imagine space being bent in any direction because our imagination isn't good enough!"

Perhaps a few analogies or models might help. Imagine that the universe is simply a vast sphere with a "skin" so thin (relatively) that we can regard it as almost *three*-dimensional (it has length, breadth, and time but *no* depth). If we imagine a large soap bubble, we have an appropriate analogy. The skin of a soap bubble is so thin that we can regard it as having virtually *no* thickness. However, were we reduced to the dimensions of mere atoms, that so-thin soap bubble skin would have considerable depth. It would, in fact, represent a four-dimensional entity (i.e., that of time included). Think now of our universe in similar vein. Embedded in the "skin" of four-dimensional space-time are the galaxies (Figure 9). There then arises an intriguing possibility. The distance in normal space-time from galaxy A to galaxy B, assuming we could make such a journey, would seem like a *straight* path so far as our senses are concerned. Due to the curvature of space, however, it would in reality be a *curved* path. There is theoretically a much shorter path by a *true* straight line. Unfortunately, this clearly involves cutting through the "skin" of four-dimensional space-time and traversing whatever lies beyond it. And that "whatever" we'll call hyperspace, the fifth dimension. Gaining access to this mysterious region is an obvious problem, and getting back into normal space-time is clearly another.

So far, as was mentioned earlier, we modeled our finite universe largely on a sphere or, more correctly, the "surface skin" of a hollow sphere. In Figure 9 for the first time we move away from the spherical model, abandoning it in favor of one regarded as physically more probable (Figure 10). The shape may seem very odd, but cosmologists now maintain that our "closed" universe is more likely to be *saddle-shaped.* It must be emphasized that this form of universe is also just a "skin." As before, without and within, lies this strange postulated fifth dimension of hyperspace.

We have spoken of a straight-line path through hyperspace between individual galaxies that are themselves within normal space-time. What of a straight-line path between *individual stars within a single galaxy?* Compared to the distances between galaxies (2.25 million light-years separate the Milky Way from its nearest galactic neighbor) the stars are located relatively close to one another. Great emphasis must be placed on the word "relatively" in this context, for the stars still lie at immense distances from one another. Since the galaxies and their stars are also within this "skin" of curved space, the same type of shortcut through hyperspace might just be possible (Figure 11). This might just tie in with the concept of a "closed" universe full of "worm holes." (See also hyperspace currents on 101).

You may justifiably ask if this strange conjectural space-geometry really helps all that much. The conventional "straight" path a normal deep-space spacecraft would require to take is in reality an arc of a circle, its length amounting to eight light-years. The shorter path through hyperspace, geometrically a chord of a circle, should prove less time consuming—perhaps something on the order of six light-years. This is not really a

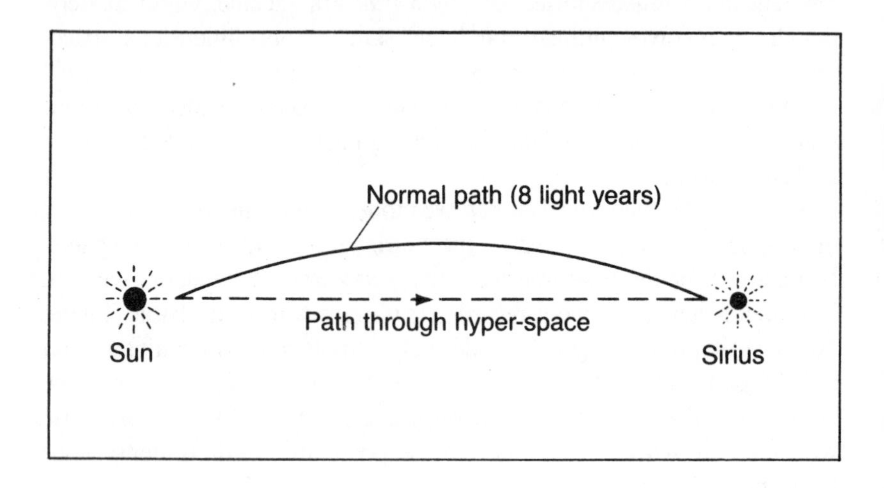

Figure 11

tremendous gain for interstellar travelers. One can hardly argue with the geometry involved. However, this is the straightforward geometry of Euclid. Two important points arise here: (1) Could this form of geometry apply in a fifth-dimensional realm?, and (2) Would time as we know it be the time of hyperspace (non-time perhaps)?

At present all this must count as speculation, but we can nevertheless ask ourselves if the journey from Sun to Sirius could be one that would occupy little or no time at all. Here we have a possible link with time travel, albeit as yet a rather tenuous one. Such unique possibilities will be further explored when we come to consider the strange geometry of black holes in chapters VIII and X.

To alter the flow of time, something we can neither see nor feel although we are acutely conscious of its passing, might at first seem ridiculous. Yet we must remember that we are dealing with a very strange universe, one that appears to become steadily stranger as cosmology continues to advance. We must be prepared to jettison ideas and concepts that we have long regarded as sacrosanct. One of the greatest cosmologists, the late Sir Arthur Eddington, acknowledged the conceptual difficulties as far back as 1931: "The theory of the expanding universe is in some respects so preposterous that we naturally hesitate to commit ourselves to it. It contains elements apparently so incredible that I feel almost an indignation that anyone should believe in it—except myself."

Now, more than half a century later, features apparently even more incredible are being suggested. These too we may eventually have to accept.

To return to the expansion of time, it is relevant to refer briefly again to what is known as time dilation at near-optical velocities. What the appropriate mathematical equations demonstrate is that at speeds approaching that of light it would be perfectly feasible to travel to another star in a period much shorter than that measured on Earth. Let us look at an extreme example. An intergalactic spaceship leaves this planet at a velocity very close to that of light, its destination the Great Galaxy in Andromeda. To its occupants a mere 27 years will pass on the journey.

Not particularly liking what they see when they get there, they decide that perhaps Earth is not such a bad place and resolve to return at the same velocity, this journey occupying another 27 years, making a total of 54 years for the round-trip. On their return they will be in for a monumental surprise, for Earth will *not* be 54 years older. It will be a planet, assuming it still exists, some *four million* years older. Since this is a daunting prospect, we might be wise to consider a much less ambitious interstellar voyage, say to the star Procyon in the constellation Canis Minor, 10.4 light-years distant, the speed of the star-ship being 0.99c again very close to that of light. To the occupants of this ship the time taken to reach Procyon will be, according to the relevant equation, 1.5 years, but to those on Earth 10.4 years will have passed. Thus a round-trip ages the occupants of the ship by 3 years but they will return to an Earth 20.8 years older. They will thus discover by the expenditure of only 3 years of their lives what life on Earth will be like nearly 21 years on! So here, theoretically at least, is a form of time travel into the not too remote future. Unfortunately, the attainment of such a fantastic velocity is totally beyond our capacity to achieve both now and for the foreseeable future. It is necessary to add that no amount of juggling with the mathematics involved can change this mode of time travel into anything but travel into the future—*never the past!*

The reader may be tempted to inquire whether or not the equation under discussion is relevant or merely a confusing mathematical paradox. So far as is known, it is perfectly valid. The concept appears ridiculous only because such high velocity values must be invoked. Whether such velocities will one day be feasible only time will tell (no pun intended).

This entire aspect is examined at considerable length in Chapter VII along with the fundamental algebra. Perhaps someday in the future we will be able to utilize the tachyon, a nuclear particle capable of *exceeding* the velocity of light. A form of "tachyon drive" then emerges as a possibility. Unfortunately, whereas a tachyon may well be capable of hyper-optical velocities, the question of mass being driven at such velocities (and there is a lot of mass in even the smallest star-ship) poses difficulties. Moreover, as we will be seeing, the insertion of a velocity greater than c into the equation brings results that, mathematically at least, make no sense. But perhaps this is just the result of limitations in our present-day knowledge and thinking.

So much then for our universe as it presently appears to cosmologists. Curved space is an odd concept, and in these last few pages we have only skimmed the subject. Our aim has simply been to present the universe as it may really be and to introduce time not merely as the aging process or something to be associated with clocks and watches. It is, in its own strange right, a dimension. We must now seek to ascertain, so far as it is possible at present, our chances of ever being able to travel through it.

Since black holes and so-called "tunnels in space" must inevitably enter into this narrative, we will be returning to the universe in due course, finding in all probability that it can be even queerer than has already been indicated in this chapter. Before doing so it might be wise to look at some of the very queer paradoxes that time travel could produce. These are intriguing and quite understandably appear to render time travel rather absurd. But, as we have seen with our universe, things are not always what they seem.

CAUSALITY AND PARADOX

One of the great, almost overwhelming difficulties inherent in time travel has been the strange paradoxes it appears to produce. Travel in space, whether interplanetary or interstellar, brings its due share of problems. Still, in a manner of speaking, these are essentially rational. Travel in time, on the other hand, by its very nature seems totally irrational. Moreover, paradoxes apart, many people think that its achievement could bring us no real advantages other than a satisfying of curiosity. Who, for example, would genuinely wish to return to the Jurassic, Triassic, or Cretaceous ages, perhaps to be chased and devoured by a ferocious, carnivorous dinosaur? And the prospect of returning to medieval Europe with its cold, unheated castles and huts, dungeons, superstition, and outbreaks of plague would surely prove attractive only to the most dedicated historians—and then only if they could be assured of a safe return to the present. And what of travel into the future, especially the far future? This, without doubt, could be highly illuminating, though in view of the course our civilization is presently taking it might be risky. Certainly there must be many of us who, contemplating the mistakes we have made, the things we should have done (or perhaps have not done) in life, wouldn't mind returning to early childhood and having a second chance. But even if we did, would events turn out differently or would life just prove to be a rerun, like the second viewing of a movie? Right away, as you'll appreciate, a paradox is rearing its head. Would we really know what mistakes to avoid, what wiser decisions to make to enable our second future to be different from the first? If so, that future could be quite different. And then, which would represent the *real* future?

Let us take a few examples of this kind totally at random.

A. A man returns to the past and for some reason or other shoots his father a year prior to his conception. He could not therefore have been born and would not exist to travel backward in time to commit the crime.

B. The parents of a young woman marry as a consequence of a chance meeting during World War II. Their daughter goes back in time, visits Germany in the fateful summer of 1939, and succeeds in assassinating Adolf Hitler. In that event World War II would have been averted. Her parents would not have met and the girl could not have been born.

C. A man travels into the future and discovers the date of his death. Two possibilities open up here. Either by traveling into the future he ages prematurely and dies on his predestined day, or his physical state is unchanged but he returns to the present knowing the precise date of his demise—hardly a comforting thought!

It is possible to draw up a virtually endless list of such strange circumstances, so where does this leave us? Time travel is regarded by many as impossible by reason of one word. That word is "causality." The dictionary defines this as "the manner in which a cause works or the relation existing between cause and effect." Neither of these definitions appears to have much bearing on our particular problem. Nevertheless, cause and effect are inextricably related. Take, for example, the case of a serious railroad accident. A rail on a section of track develops a fault. This represents a *cause.* The *result* is a catastrophic pile-up. The result was the consequence of the cause. Our normal day-to-day experience allows us no other conclusion. Cause *precedes* effect, and it is absolutely certain that during the course of our lives we will experience no violation of this. The implication, therefore, is that since travel in time negates this, it must as a consequence be impossible.

Some twenty years ago the British Broadcasting Corporation produced an interesting TV series entitled *Time Tunnel*.* This was, of course, pure science fiction—and perhaps not of the best kind. Two particular episodes have remained fixed in my memory. I should mention that the "Time Tunnel" was simply a large aperture, like an outsize picture frame. Whoever stepped into it was immediately taken back in time to a period, and certainly under circumstances, not of his or her own choosing. In one episode an unfortunate time traveler finds himself on board the speeding White Star liner *Titanic* on the evening of April 14, 1912. Now, as everyone knows, this great vessel collided a few hours later with an iceberg that ripped open her starboard side like a tin can for 300 feet beneath the waterline. Early the following morning she sank with a loss of over 1,700 lives—the worst peacetime maritime disaster in history. Of course the time traveler, well aware of what was coming, did his best to alert captain and crew to the impending tragedy—without success. Had he been successful, the disaster would not have occurred—but as everyone knows, it did. Another intrepid time traveler later in the series finds herself in the vicinity of the already-erupting volcano Krakatoa, which, as history records, blew up with cataclysmic effect on August 27, 1883, producing the greatest recorded bang in history. (It was heard on Rodriguez Island in the Indian Ocean nearly 4,000 miles distant.) The time traveler does her best to warn all those in the vicinity to distance themselves as far and as quickly as possible from the volcano. Of course they don't, and the death toll was horrific. The girl knew this, but had her efforts met with success, the future she knew would not have occurred—in which case it would not have been the future. Going back in time clearly prevents our interfering with what is apparently predestined.

Admittedly, it is extremely difficult to find a way around this kind of impasse so long as we regard time in terms of a one-way ongoing stream. For a moment or two let us consider a crude analogy. A phonograph record is a disc not of concentric grooves but of one long spiral. This spiral

*In the late 1980s American television viewers were introduced to a time travel series entitled *Quantum Leap*.

begins near the perimeter and ends close to the center. Like time, as we generally understand it, this is an ongoing route. The stylus of the record player starts at the edge of the disc and terminates near the center unless some jerk or fault causes it to jump back to an earlier part of the spiral. Something of what we have heard already we will therefore hear again. Let us for a moment accept that time is like that. As I write these words could it be that the "spiral" of time we occupy today runs very close in some strange dimension to a part that represents yesterday and equally close to another that represents tomorrow? But what conceivable kind of "jerk" could cause our particular "life needle" to jump a groove either way? Certainly none that can be visualized.

Let's return to the man who, during a time trip into the past, killed his father before he had even been conceived. Since he *does* exist, he must have been conceived and born, and therefore we must assume *either* that the murder of his father could not have taken place or that something prevented its occurrence. Perhaps he shot the wrong man in error. Such

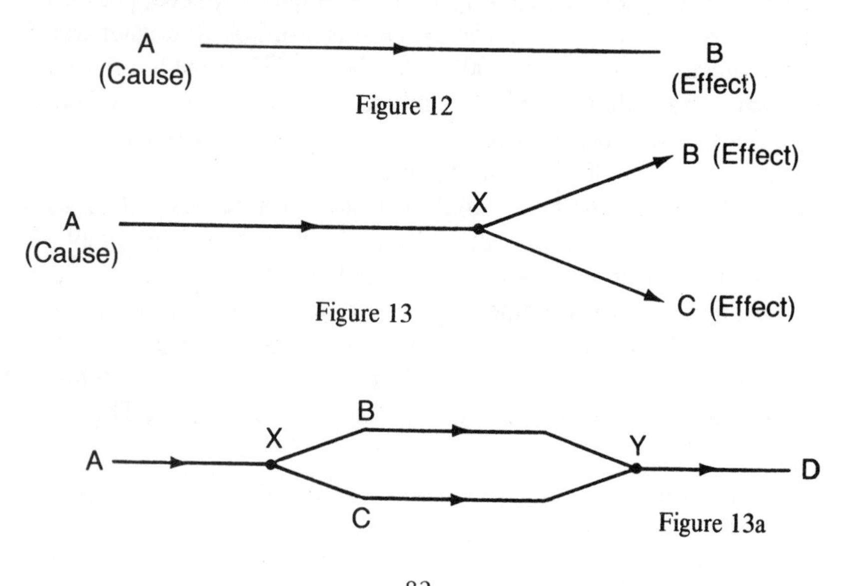

Figure 12

Figure 13

Figure 13a

explanations are totally invalid, for they dodge the issue entirely. The paradox is still with us. Can it possibly be resolved?

Suppose the man's father is both killed and not killed, in which case his son exists and yet does not exist. Surely this is ridiculous. It must be one or the other. Common sense tells us that an event either takes place or it does not. No doubt this is perfectly true if time is just a simple, onward-flowing stream. Consider, however, two roads. One is perfectly straight (Figure 12); the other contains a forked junction (Figure 13). These might be considered as alternative forms of time road.

In Figure 12 causality is not violated since a cause at A can lead only to the effect at B. This represents the passage of time as we have always visualized it. In Figure 13, however, it would appear that a cause at A could have two possible effects, B or C, depending on which time branch is followed when the point of divergence X is reached. In Figure 13a, a somewhat similar state of affairs prevails save that the two branches reunite at Y.

Should a time stream have evolved in this unorthodox manner (at least it seems unorthodox to us), it would imply that when a person's life reached the point of divergence X in Figure 13, in some strange way that we cannot visualize, an exact facsimile of that person could begin to exist in a different but parallel time stream performing actions that were totally different but amid identical surroundings. Meantime, for the sake of argument, we must assume that on reaching point X that person would be unaware of what had happened so that both he and his "twin" continued their lives quite unperturbed.

All this should be seen rather as a model than as actuality. Later we will see how contemporary physics develops (notably with respect to black holes and "parallel" universes) a potential for explaining away some very odd complications. As the Bible puts it, "we see through a glass darkly." Perhaps human beings are beginning vaguely to understand. What has just been postulated might have real and practical significance.

No one would attempt to deny that visualizing time along these strange lines involves immense conceptual difficulties, and to some of us such ideas must seem like a downright affront to reason and common sense. Yet if time travel is ever to have any relevance, it is essential we come to grips with totally new concepts regarding the nature of time itself. We must unshackle our thinking.

The idea of parallel or branched time streams is by no means novel to writers of science fiction, and over the years several writers have developed it. One of the best examples is *Bring the Jubilee* by Ward Moore. The main character exists in a world the history of which is that of ours up to the American Civil War. Thereafter time (and history) diverge, and he enters a world in which the war was won not by the Union but by the Confederacy. The reader is then treated to an exposition on the kind of society that might have developed in the aftermath of a Confederate victory. The hero of the story then achieves a capacity for time travel, goes back to the wartime years but on this occasion is switched to the time track in which the North wins. Unable to return in time to the victorious Confederacy, he finds himself in circumstances totally distinct from those he once knew and loved. He must now ponder several vital questions. Which is the real world, and what has happened to that other with which he was once so familiar? Was it a dream, did it ever really exist, and if so is it still there running along on a parallel time track? Figure 13a illustrates the position, although the geometry of this diagram allows a convergence of the two loops. No such convenient circumstance is permitted to the "lost" time traveler short of returning to point X and trusting he will be switched to the Confederacy "stream."

Despite what has been said so far regarding time branching it is only fair to state, as we hinted earlier, that in the light of current scientific opinion genuine travel into future, past, or sideways is much more likely to have a cosmological basis. In other words, space travel in one form or another must also be involved. Indeed, this was the reason for reviewing our knowledge and thoughts regarding the universe in the preceding chapters. The concept of loops and branches represents a much more philosophical approach to the problem, although this certainly does not mean it should be disregarded; some scientists refute very vigorously the idea that time can be compared to an ever-rolling stream. In fact a leading British cosmologist, Sir Fred Hoyle, is on record as describing it as "a grotesque and absurd illusion." He amplifies this by stating that *all* of time has *equal* validity and that what we choose to term the present is not of particular importance. He contends that in the four-dimensional realm of space-time the *entire* history of our planet is laid out as a spiral in four dimensions moving both around the Sun *and* through time. To quote Hoyle again, "There can be no question of singling out a special point in the spiral and stating that this particular point represents the present position of the Earth." This must lead us to wonder why the present is apparently a special point in this strange four-dimensional spiral. The answer is that it isn't—it is only the inherent conventionality and restriction in our thinking that lead us to pose this question. The analogy made earlier with a phonograph record may be of some assistance here.

If you felt you were being invited to explore a strange, uncharted river system when the concept of time loops and time branches was introduced, you must, quite understandably, now feel that you are being led into an immense, unknown ocean where apparently nothing is as it seems and where common sense is only a hindrance. In these circumstances it is best to try for the moment to lay aside conventional ideas regarding space and time—admittedly no easy task since these are the inevitable products of a lifetime's practical experience.

Hoyle's basic contention is that *everything* exists, and this must include everything that *was* and everything that ever *will be.* Back to that phonograph record for a moment or two. The point the stylus has reached on its surface is the *present,* the area already covered is the *past,* and that yet to be traversed is the *future.* The entire disc represents a four-dimensional spiral in which it might be said that, in the broadest terms, past, present, and future are all there in a single entity, although this cannot represent a total analogy with Hoyle. How then can his contention be justified? He has stated that everything exists that ever was or ever will be, but surely objects or beings that once existed but were destroyed cannot still exist? During the Jurassic, Triassic, and Cretaceous ages, for example, our planet was dominated by giant reptiles. At the close of the Cretaceous these were suddenly wiped out, it is thought, by the effects of an asteroid colliding with Earth. They cannot therefore still exist. But is this not an example of restrictive thinking due to convention? The dinosaurs do not exist *in the present,* but they still dominate Earth *in the past.* Past, present, and future do not exist together. They cannot. Neither do the points representing past, present, and future on that phonograph record. But that record is an entity, a complete thing. So also is the four-dimensional spiral of space-time. The essential difference is that we can see and handle the record, all of it, in the present. The only portion of the four-dimensional spiral of space-time we can possibly experience is the present. We remember the past; of the future, as yet, we know absolutely nothing.

If we accept Hoyle's contention, an interesting point emerges. In the phonograph disc it is obvious that what lies ahead of the stylus traveling in the spiral groove is fixed, preordained. Nothing is going to alter that. Are we entitled to assume that in this sense the spiral disc of space-time is analogous, that our futures are also preordained? We might think not, that we could take some action that would certainly change the course of our future such, for example, as a decision to go and reside in another country. But would we really be changing our future? Would we not just have reached that point in the space-time parallel where this particular decision had been preordained?

Despite the bizarre nature of Hoyle's belief it is clearly at variance with the concept of diverging and parallel time streams. Neither is it in strict accord with the idea of time as an ever-rolling stream if past, present, and future are all, in their way, integral parts. His ideas rather neatly link us with the cosmic concept via a philosophical bridge. It is the cosmic aspect with which we will primarily be dealing in the ensuing chapters. We will be entering the strangest of strange universes, one far removed from that detailed in orthodox astronomy textbooks wherein galaxies, stars, planets, moons, asteroids, comets, and meteors have their allotted places and courses in a simple great immensity we term space. Einstein and others have changed the seemingly understandable well-ordered universe into something totally different. It is one that should interest and intrigue, not alarm. It is most unlikely that this stranger universe surrounding us is ever going to change *our* lives in any way, although it may well do so for our remote descendants. Meantime we can still enjoy the splendor of the conventional heavens, still admire the magnificent Orion sprawled across the winter sky, watch with delight the familiar little cluster of the Pleiades rising in the east, or Sirius on a clear, moonless frosty night flashing brilliantly low in the south. Nevertheless we must always remember that space must be as it *is* and *not* as our tidy, conventional minds feel that it ought to be.

I think I should emphasize again that most if not all of the concepts about to be discussed in the ensuing chapters will seem, especially at first, both unreal and illogical. "Parallel" universes, "tunnels" in space-time, time warping, and the like are not ideas likely to occur naturally. Even to those who, like myself, have had an active interest in astronomy covering many years, the new concepts had at first a distinct air of unreality about them. My devotion to astronomy started in January 1938, some two weeks prior to my fifteenth birthday. The accepted cosmological outlook on the universe then was considerably different from what it is today, yet the stars then are the stars of the present. The heavens, even through a telescope, look just as they did in 1938; the universe has not really changed although man-made orbiters and satellites now pass across these same familiar constellations. It is ideas that have changed, and as a consequence we have come that much closer to the fundamental truth though a long tortuous road still lies ahead of us.

TRAVEL INTO THE FUTURE

Time travel involves leaps backward or forward in time. What must be of greatest interest to us, of course, are jumps of considerable length. Any technique that merely takes us back or forward a few days would, for the most part, be only of academic interest though its potential toward greater things cannot be denied. We must remember that the frail flying machine of the Wright brothers led in only a few decades to the *Concorde,* the 747, and the DC-10. Personally, I have no particular desire to return to yesterday or the day before that because, so far as I am concerned, there was nothing particularly special about either. Neither am I particularly interested about a jump into tomorrow or the day after—each will come soon enough in any case. Admittedly, a distinct financial advantage might accrue from knowing two days in advance the winner of the Kentucky Derby—so long, that is, as the technique involved permitted the individual concerned to safe return in time to place his bet!

In choosing a title for this chapter I almost decided upon "Elastic Space," but since the theme is time travel into the future I felt this would probably be inappropriate. Elasticity is a term we normally associate with such everyday articles as rubber, gum, stretch fabric, and even springs. Applied to phenomena such as time and space, the notions associated with elasticity aren't quite adequate. Again we are face to face with the consequence of rational and restricted thinking. We tend, quite naturally, to believe only what we see. For many centuries men and women thought the Earth was flat simply because it looked flat—the horizon was merely a perspective effect beyond which presumably lay still more flat Earth. Since the true nature of gravity was not then understood, the unchallenged belief was that Earth must have a boundary somewhere; any mariners ill-advised enough to sail too far beyond the horizon would inevitably perish as their ships tumbled off the edge of the world. They also believed implicitly that the Sun, and all the rest of the universe, revolved around the Earth simply because it looked that way. Copernicus and Galileo stirred up a real hornet's nest when they had the temerity to suggest that, far from being the center of things, Earth was merely one of a number of relatively minor bodies that orbited the Sun, the real center of the Solar System. Old ideas tend to die hard when they appear to contradict common sense. (I am reliably informed that even now there exists an august body known as the Flat Earth Society!)

To us such notions seem incredibly quaint, the more so since human beings have begun to explore space. The world may look flat, but there can be no doubt whatsoever that it is round. Neither do we doubt that Earth orbits the Sun and not vice versa, however it may appear. We have come to terms with these things despite the fact that such truths were rank heresy to our ancestors of a few centuries back. Thanks to Newton, the basic precepts of the Solar System are freely accepted. Our forebears thought dropped objects fell to the ground simply because there was nothing to hold them up (my personal fixed belief as a small boy). It took Newton to sort this one out, a train of thought allegedly initiated by his seeing an apple fall from a tree (some say it fell on his head).

Reluctance to accept what the mathematics of relativity shows begins, I think, in the minds of most people when they are first confronted by the famous "clock paradox" that was mentioned briefly in Chapter IV. This can be summed up as the apparent dilation or "slowing down" of time as recorded by a chronometer on a star-ship moving at a high sub-multiple of the velocity of light, compared to that measured by a stationary clock on Earth. Surely, it seems, a clock in motion cannot possibly behave differently from one that is stationary. Admittedly, I have no reason to believe that the clock in our car differs in its reckoning of time from the watch on my wrist when my wife or daughter is driving that car at 60 miles per hour while I am sitting at home. It is only when high, relativistic velocities come into the equation (i.e., speeds close to that of light) that the effects begin to manifest themselves. Time then becomes dilated to such a degree that really odd things start to happen. Time in these circumstances does begin to display a measure of elasticity inasmuch as it is being "stretched."

In his Special Theory of Relativity, Einstein described the "clock paradox" in the following terms: "If we were to place a living organism in a box one could arrange that the organism, after an arbitrarily lengthy flight, could be returned to its original spot in a scarcely altered condition while corresponding organisms which had remained in their original positions had long since given way to new generations. So far as the moving organism was concerned the lengthy time of the journey was a mere instant, provided the motion took place with almost the speed of light."

An analogous effect would occur with respect to space, manifesting itself with respect to material objects such, for example, as a piece of steel rod of known length. To the occupants of the hurtling star-ship, as with the clocks, nothing would seem untoward. Compared to an identical piece of steel rod stationary on Earth, the length of that space-borne piece of rod would decrease—i.e., as velocity increases, length decreases.

$$L_v = L_s (\sqrt{1 - v^2/c^2})$$

where,

L_v = length of object moving at velocity v
L_s = length of object at rest

As v tends toward c then

$$\sqrt{1 - v^2/c^2} \text{ becomes very small.}$$

therefore

$$L_s(\sqrt{1 - v^2/c^2} \text{ becomes very small.}$$

therefore

$$L_v \text{ becomes very small.}$$

At $v = c$ the right-hand side of the equation equals zero.
Then, L_v = zero, a nonsensical result demonstrating once again that the attainment of light velocity is impossible.

To take this aspect to its logical conclusion we must consider what would happen to a star-ship itself were it to achieve the speed of light. It would then have *infinite mass,* which as we saw in Chapter IV is an equally absurd result leading to the same conclusion.

Suppose a spacecraft moves with velocity V. If that velocity is increased by an increment U then the overall velocity becomes $(V + U)$. By relativistic laws the resultant of such an addition, which we'll call W, can be expressed algebraically as follows:

$$W = \frac{V + U}{1 + \frac{UV}{c^2}}$$

where c is the velocity of light.

If V equals 0.9c and $U = 0.1c$, it would appear that the final velocity should be a c (that of light). This, however, is *not* the case. If we substitute in the foregoing equation, then:

$$W = \frac{0.9c + 0.1c}{1 + \frac{0.1c \times 0.9c}{c^2}} = \frac{c}{1 + \frac{0.09_c}{c^2}} = \frac{c}{1.09} = 0.918c$$

This is close to the speed of light and is a tremendous velocity, but it is *not* c. Similarly if we make the increment U equal to 0.2c, the resultant is not 1.1c but 0.923c. We have come even closer to the speed of light but have still not attained it. It may seem that we will eventually attain it by the addition of sufficient increments, but things just don't work out that way. If we add nine increments of 0.1c, the effective velocity is not doubled to 1.8c but remains at 0.944c. Surely, were we to add ten such increments, the final value must be c. This is invalid, for we must remember that ten such increments would themselves constitute light velocity.

$$W = \frac{0.9c + 1.0c}{1 + \frac{0.9c \times 1.0c}{c^2}} = \frac{1.9c}{1.9} = c \text{ (invalid result)}$$

So thus, irrespective of the velocity increments we add, the space vessel can *never* attain the absolute velocity $W = c$, although, theoretically at least, we could certainly get fairly close to it.

It might be appropriate at this point to refer briefly again to tachyons—subnuclear particles which, if they really do exist (and there is now good reason to believe they do), are thought to travel at velocities greatly in excess of that of light. This would appear to make complete nonsense of the above equation and bring into serious doubt the validity of the relativistic law upon which the results are based. Tachyons, incidentally, when approaching the velocity of light are in fact *slowing down.* For interest's sake, let's insert such a hyper-optical velocity into the equation and see what results. Suppose V is equal to 3c (due to some form of "tachyon drive") and U is equal to 0.5c. As before:

$$W = \frac{3.0c + 0.5c}{1 + \frac{3.0c \times 0.5c}{c^2}} = \frac{3.5c}{1 + 1.5} = \frac{3.5c}{2.5} = 1.4c$$

This is the kind of result we might have anticipated. There is, however, nothing wrong with the equation. We should at this point return to the equation involving mass, an equation we used in Chapter IV. From this it was clear that at light velocity mass becomes infinite. A space vessel has mass and plenty of it, and for this reason is unable to achieve light velocity. It is when something that possesses *no* mass is involved that the position is so radically changed. Since tachyons are not considered to have mass, the problem does not arise.

The faster anything having mass travels, the slower does time pass. If the velocity of light *were* attainable, time would simply stand still. Since tachyons are capable of velocities greatly in excess of that of light, we are faced with a most intriguing possibility: time must run in reverse! In other words, backward into the past. In a subsequent chapter we will be taking a close look at time travel into the past, so here we merely mention the fact and will defer a closer examination until then.

Meantime let us return to this odd paradox of time running more slowly and its implications for travel into the future. It might, as a convenient starting point, be useful to employ an analogy. If we are traveling at 70 miles per hour along a straight track in a railroad train and another train on a strictly parallel track overtakes ours, draws level, then regulates its speed exactly to that at which we are traveling, we immediately become conscious of a curious effect. Looking out the opposite windows, we see the countryside still flashing past. But looking toward the other train, which is effectively obscuring the scenery on that side, we seem to have become stationary despite the sound of the wheels on the track and the motion of the train. This is a perfectly normal phenomenon with which no doubt many of us are familiar. The explanation is that we now have *two* frames of reference. Relative to the stationary countryside we are traveling at 70 miles per hour, but relative to the other train we are *at rest*. Let us now translate this into more precise relativistic terms.

If we could move with the velocity of light, we would be stationary with respect to all other objects traveling at the same velocity. There is one proviso, however—we must move in a *straight line* in order to keep our velocity constant. That is why I stressed in the above analogy that the two trains were on parallel, straight tracks. If the track on which the other train was traveling was less than parallel, we would soon be aware that our train was no longer stationary with respect to the other.

Two frames of reference are also invoked in the case of a star-ship traveling at near-light velocity and the Earth it has left behind. At such a velocity those aboard the star-ship would be unaware that their clocks and watches were running so much more slowly and that the dimensions of their ship had changed. If, however, it were feasible for them to observe a clock on Earth, they would become immediately aware of the bizarre disparity that had occurred. It was shown earlier that a star-ship which had somehow attained the velocity of light would possess infinite mass since mass increases with velocity. Since infinite mass is a clearly absurd proposition, we must assume that the attainment of light velocity or beyond is quite impossible for space vehicles. But if in some miraculous manner it could be attained, the clocks on the ship would *stop* relative to those on Earth. In other words, though the interval between successive ticks were to the star-ship crew only normal seconds, one of these seconds to us on Earth would be infinite. Time would have become frozen!

This slowing down of time is merely the logical consequence of velocities approaching ever closer (but never reaching) that of light. Since light velocity cannot be attained, the ship's clocks will not stop so far as we are concerned, but they will appear to run ever more slowly compared to our own. This then is what is termed time dilation.

Now perhaps for the first time we can begin to appreciate why such a vessel moving at the merest fraction less than light velocity (.9999999999c) could traverse the 4.5 million light-years of a round-trip to the Great Galaxy in Andromeda (2.25 million light-years remote from Earth) in the absurdly short time of around 55 years and arrive back to an Earth some 4.7 *millions* of years in the future. Admittedly, the men and women of the crew, in their twenties when they departed, would return as elderly men and women of 75 or thereabouts. Most of their lives would have been spent on the journey there and back, but at least it would have been a feasible proposition—one of 4.7 million years assuredly would not. This would indeed be time travel into the future in the fullest possible sense.

Such a journey represents an extreme example, but we can now see something of the relativistic advantages for interstellar travel which near-light speeds could achieve. From a practical point of view we are bound to remark that the energy requirements would be "astronomical." For a spaceship weighing only *one ton* capable of a velocity of 240,000 kilometers per second (150,000 miles per second) or 0.8c, these requirements are estimated at around 215 billion kilowatt-hours—about as much as is generated by *all* the power stations on Earth working flat out over a period of several months. Indeed, a "time machine" based on this technique would have a truly insatiable appetite. The problem of energy is therefore a highly daunting one, though any such calculations ignore what might be possible if some sort of matter/antimatter drive became practical. But one vital problem we cannot ignore is the fact that every tiny speck of cosmic dust could prove lethal to the travelers were it to impact their spacecraft moving at such fantastic velocities. If it becomes possible to create vessels capable of these sort of velocities, it would not, we trust, be beyond the power of space engineers to devise some form of shielding or deflecting device.

Is it possible, you may be tempted to inquire, to prove experimentally that the attainment of such fantastic velocities would "slow" time? The answer is yes, although practical proof cannot at present be provided by dispatching astronauts into interstellar space on a round-trip to a star at near-light velocities. Confirmation comes as a result of research into the nature of the elementary particles that constitute cosmic rays. In several instances these have been found to travel very close to the speed of light. (For full details, see Appendix 1.)

Cosmic rays reach Earth from outer space. The particles are highly energetic and continually bombard the atmosphere of our planet. They are primarily nuclei of the most abundant elements, although all known naturally occuring nuclei are represented. These are termed *primary cosmic rays.* On entering our atmosphere they collide violently with atomic nuclei *in* that atmosphere, thereby producing *secondary cosmic rays,* which consist of elementary particles. At low altitudes a certain proportion of cosmic rays are identified as particles known as micro-mu mesons. The mass of a micro-mu meson, though slightly more than two hundred times that of an electron, is still so incredibly slight as to be negligible. They may travel at a low speed, or nearly as fast as light. It is a fundamental feature of micro-mu mesons that they are highly unstable and decay easily into other particles. The time required for this decay to take place has been measured by means of a relativistic equation relating speed and time. This equation is expressed as the ratio of the life-rate as measured by a clock related to a micro-mu meson as against a clock related to the observer. The formula is as follows:

$$T_{Earth} = \frac{\sqrt{1.53 \text{ microsecs.}}}{1 - \frac{v^2}{c^2}}$$

The upper term in the right-hand side of the equation represents the life-rate of a *stationary* micro-mu meson, 1.53 microseconds, which is a very short period indeed. From this it can be shown that a micro-mu meson in motion at high velocity V exists for a longer period of time than does one at rest. This follows from simple algebra. If V is large, so must also be V^2. It follows therefore that the ratio

$$\frac{v^2}{c^2}$$

tends toward (but does not reach) unity. This being so, then

$$1 - \frac{v^2}{c^2}$$

becomes a very small quantity, and consequently the entire right-hand side of the equation has a higher value. T_{Earth} being equal to this is therefore also higher. Put more simply, the period in which a micro-mu meson in motion at high velocity exists is greater than for one at rest.

It is possible to generate beams of micro-mu mesons having more or less the same velocity; the life-rate of these has been measured as a function of their velocity. The greater the velocity of the beam the greater its life-span and vice versa. Thus a fundamental proposition of relativity has been shown to be correct.

As might be expected the similarity of the above equation with that used to estimate the extent of time dilation as recorded by clocks on a star-ship moving at a high submultiple of light velocity compared to those on Earth is very close. This is as follows:

$$\frac{T_{ss}}{T_t} = \sqrt{1 - \frac{V^2}{c^2}} \quad \text{or} \quad T_t = \frac{T_{ss}}{\sqrt{1 - \frac{V^2}{c^2}}}$$

where T_{ss} is the time as measured aboard the star-ship; T_t is the time as measured stationary on Earth; V is the velocity of the star-ship; and c, as usual, represents the velocity of light.

A few specific examples are probably in order at this point. Since the star closest to us is Proxima Centauri (the binary companion of Alpha Centauri), 4.2 light-years distant, this would seem to be the obvious first destination. If the ship has a velocity of 0.66c or 200,000 kilometers per second (two-thirds the velocity of light), the result is as follows using the well-known equation:

$$\frac{T_{ss}}{T_t} = \sqrt{1 - \frac{V^2}{c^2}}$$

The reasoning is as follows:

a. Distance equals velocity multiplied by time, i.e., $D = VT$

b. Time equals distance divided by velocity, i.e.,

$$T = \frac{D}{V}$$

c. The velocity of light c is 1 light-year per year, i.e.,

$$c = \frac{\text{1 light-year}}{\text{year}}$$

d. The velocity of the star-ship is

$$\frac{2}{3}c$$

e. The distance to Proxima Centauri and back is 8.4 light-years, i.e., D = 8.4 light-years

Using the above information the calculation is as follows:

$$T_t = \frac{D}{V} \text{ i.e. } T_t = \frac{\text{8.4 light-years}}{\frac{2}{3}c} = \frac{\text{8.4 light-years}}{\frac{\text{2/3 light-year}}{\text{year}}}$$

$$\therefore T_t = \text{8.4 light-years}\left(\frac{\text{year}}{\text{2/3 light-year}}\right) = \frac{\text{8.4 years}}{2/3} =$$

$$8.4 \times 3/2 \text{ years} = 12.6 \text{ years}$$

$$\text{Thus } \frac{T_{ss}}{T_t} = \sqrt{1 - \frac{(0.66c)^1}{c^2}} = \sqrt{1 - \frac{0.436c^2}{c^2}} = \sqrt{0.564} = 0.751$$

$$\therefore T_{ss} = 0.751\ T_t = 0.751 \times 12.6 = 9.46 \text{ years}$$

The net result for those aboard the star-ship would be a savings of (12.6 – 9.46) years, or 3.14 years. Allowing for periods of course correction, acceleration, and deceleration, we might round this off to 3 years. Admittedly, this is hardly a massive time jump into the future and to achieve it a velocity two-thirds that of light has been necessary as well as a journey to and from a star, albeit the one lying closest to the Solar System. Nevertheless, the crux of the matter is this: travel into the future is theoretically possible.

Now let us consider a somewhat larger jump into the future. The above calculation tells us that this will necessitate a round-trip to a more remote star and at a velocity greatly in excess of 0.66c, tremendous though that velocity is. On this occasion we will select the lovely winter star Procyon in the constellation of Canis Minor, the Little Dog. Reasoning as before:

a. Distance from the Sun = 10.4 light-years
b. Distance of round-trip = 20.8 light-years
c. Velocity of star-ship = 0.99c or

$$\frac{99}{100}c$$

$$T_t = \frac{D}{V} = \frac{20.8 \text{ light-years}}{0.99\text{ c}} = \frac{20.8 \text{ light-years}}{\frac{0.99 \text{ light-year}}{\text{year}}}$$

$$\therefore T_t = 20.8 \text{ light-years}\left(\frac{\text{year}}{0.99 \text{ light-year}}\right) = \frac{20.8 \text{ years}}{0.99} = \frac{20.8 \text{ years}}{99/100}$$

$$\therefore T_t = 20.8 \times \frac{100}{99} \text{ years} = \frac{2080 \text{ years}}{99} = 21.01 \text{ years}$$

Using the time dilation formula, the time aboard the star-ship is 2.94 years. A period of nearly 21 years on Earth has been reduced to nearly 3 years. In other words, returning space voyagers have aged about 18 years less than persons on Earth. Again, course changes, plus periods of acceleration and deceleration, would reduce the advantage somewhat, so 15 to 16 years might represent a more realistic figure. This is a reasonable but still not gigantic jump into the future, and of course the power requirements necessary to achieve 0.99c would be gargantuan by any standards.

By selecting still more distant stars the time differential could obviously be increased. A 540-light-year journey to the beautiful blue-white star Rigel, in Orion at 0.66c would to those on Earth take nearly 1500 years. It would take only about 1100 to those on the star-ship. Since both periods vastly exceed that of the human life span, further comment is unnecessary. At distances of this order and greater, only a velocity of 0.99c will suffice. And even that would prove insufficient for the more remote destinations.

This seems a convenient time to mention more bizarre concepts that might conceivably permit rapid transit between stars and so also provide us with possible leaps into the future.

It has been suggested by a few eminent and reputable scientists that every region of the universe is interconnected with every other by fast "hyperspace" currents that flow between every possible departure and arrival point in space. Noted aerospace scientist Alan Hold has even postulated that almost instantaneous interstellar transit might be achieved by "riding" the hyperspace currents that flow from star to star. This, he believes, could be accomplished by generating coherent patterns of electromagnetic energy that would interact with these hyperspace energy currents, thereby reducing or augmenting the gravitational field around the ship. These might be generated in a toroidal-shaped configuration, the process employing such devices as high-energy lasers, superconducting magnets, and electron beams. If such a concept can only be described as mind-boggling, presumably so would our journeys to the Moon have seemed only a century ago. Discussion of this technique might have been more relevant to our examination of journeys through a fifth dimension, with which we have already dealt, or voyages via black holes, a subject with which we will be dealing shortly. In any case it is at the very least an appropriate and intriguing digression.

And most readers still may be unsatisfied with any explanation involving the concept of infinite mass. Briefly, the position is that a body having infinite mass (e.g., a star-ship moving at c, the velocity of light) could only have been accelerated to that speed by an *infinite force,* something that is clearly ridiculous. This seems another proof that the attainment of light velocity is impossible. What then of those previously mentioned tachyons and the possibility of hyper-light "tachyon drives"? Such phenomena and their relation to relativistic space/time travel cannot be arbitrarily dismissed; we'll look at the reasons when we deal (in Chapter X) with time travel into the past.

For the moment, however, we'll confine our attentions to time travel into the future by the use of near-light velocities. You'll recall that relativistic travel (assuming that ship and crew return safely to Earth to observe the effect) can be *only* into the future and not the past. The mathematics render this very clear. Unfortunately, therefore, the optimistic young woman described in the following limerick won't be able to follow the timetable laid out for her trip—not, that is, unless she knows something we don't.

There was a young lady named Bright,
Who traveled much faster than light,
She started one day
In a relative way,
And returned on the previous night!

Whatever the prospects of going back in time, it is hardly likely to be achieved by time dilation due to the attainment of near-light velocities.

Time dilation has certainly negated the conventional concept of time as a leisurely-flowing stream. We see time now for what it really is—a physical dimension that, like the other more conventional dimensions, can be altered. The faster we travel the more slowly does time pass, but the main hindrance, as has already been emphasized, is that the sublight velocities required are so very high.

A number of years ago in my book *Journey to Alpha Centauri* I gave my star-ship, the *Columbus*, a velocity of 0.02c, that is, one-fiftieth the speed of light. Since this is 3,720 miles per second or 13,392,000 miles per hour, it can hardly be regarded as a crawl, yet even this is well beyond our present capacity to achieve. Despite such a velocity, the transit time to Alpha Centauri, the star closest to the Sun, amounted to some 215 years. Seven generations of men and women lived and died on *Columbus* before Alpha Centauri (or, rather, a suitable planet of it) was reached. Had a return journey been involved, a further 215 years would have proved necessary. Such a relatively low velocity is therefore quite useless in the context of travel into the future.

Though we are presented with a theoretical but as yet impractical method of time travel into the future, we can already see another paradox. A journey at high velocity into interstellar space could justifiably lead those on Earth to maintain that the clocks on the star-ship were simply running *slow*. The occupants of the star-ship could claim with equal emphasis that clocks on Earth were running *fast*. All depends on the particular frame of reference: star-ship moving, Earth at rest; or star-ship stationary, Earth moving. This paradox is one that is easily resolved for it *is* the star-ship that is moving relative to Earth and not vice versa, although certainly to the star-ship occupants their vessel might seem at rest against the distant stellar background with Earth apparently receding. The aging of those friends left behind on Earth compared to those on the returning star-ship would presumably remove any element of doubt as to the true state of affairs.

By traveling at relativistic velocities we are simply stretching time. We could put it another way, however, and say that space is being compressed. Table 1 will give you some idea of the extent by which time dilation can effectively shorten the time of transit to a number of stars. If each journey is to be a round-trip to the star and back to Earth, the figure in the "Time Saving (years)" column is doubled for the "Time into future (years)" column. The latter, of course, indicates the amount of time the star-traveler has advanced into the future relative to Earth. The table is based upon

TABLE 1
(Applicable to star-ships at 0.66c)

Star	Earth years	Star-ship years	Distance (light-years)	Time Saving (years)	Time into future (years)
Canopus	1733.0	1302.6	650.0	431.0	n/a
Vega*	69.3	52.1	26.0	17.2	34.4
Capella*	125.3	94.1	47.0	31.2	62.4
Arcturus*	109.3	82.1	41.0	27.2	54.4
Rigel	1440.0	1081.9	540.0	358.1	n/a
Aldebaran	152.0	114.2	57.0	37.8	n/a
Spica	613.3	460.8	230.0	152.5	n/a
Antares	960.0	721.0	360.0	240.0	n/a
Altair*	42.6	32.0	16.0	10.6	21.2
Pollux*	85.2	64.0	32.0	21.2	42.4
Castor*	114.6	86.1	43.0	28.5	57.0
Regulus	149.3	112.2	56.0	37.1	n/a
Close stars					
Tau Ceti*	27.2	20.4	10.2	6.8	13.6
Epsilon Eridani*	28.0	21.0	10.5	7.0	14.0
61 Cygni*	28.5	21.4	10.7	7.1	14.2
Epsilon Indi*	30.9	23.2	11.6	7.7	15.4
Wolf 359*	21.6	16.2	8.1	5.4	10.8

a velocity of 0.66c (two-thirds the velocity of light) which, as we saw earlier, is barely adequate. Only those stars marked with an asterisk can realistically be regarded as sufficiently close for a jump into the future at 0.66c. That is, of course, unless the average human life span can be extended much beyond its current length. It must be emphasized that all this has referred to motion at *constant* velocity. Yet from a practical viewpoint we accept that the star-ships we have been envisaging must have periods of acceleration enabling them to attain the velocity of 0.66c, 0.99c, or whatever, and similarly that they must have like periods of deceleration as they approach their objectives. This means that the figures in the tables are a bit low, but on the whole, these periods of "stepping on the gas" and "applying the brakes" would be *relatively* brief compared to the journeys the travelers would make at *constant* velocity. Course changes, if necessary, would further slightly reduce the time advantage.

Readers who are devotees of science fiction will be familiar with the term "time warp." Followers of the adventures of Captain Kirk and the crew of the star-ship USS *Enterprise* in the popular TV series *Star Trek* will be familiar with a similar term. At the command "Warp factor two" or the like, the *Enterprise* proceeds "to boldly go" (surely the best-known split infinitive of all time) "where no man has gone before." (Kirk's latter-day successor in command, Captain Picard, ventures where "no *one* has gone before.") Though "warp drive" may seem just a convenient cliché to explain the inexplicable, the notion has at least some underlying substance. However, what relation Captain Kirk's "warp factors" bear to true time warps is more than a little vague, for neither he nor any of his crew ever seem to have problems with time dilation.

A true time warp is the consequence of an intense gravitational field. In its immediate environs the fabric of space-time becomes considerably distorted. What would be the effect on a star-ship clock were such a vessel, either by accident or design, to enter such a field? Would it be analogous to what happens to a clock on a star-ship traveling at a relativistic velocity? The answer is yes, that clock, too, would run slow. A short while back we considered what would happen aboard and to a star-ship that in some miraculous manner managed to attain the speed of light. Between every tick of a clock aboard the vessel time would be infinite. So far as those on Earth were concerned there would be no more ticks; nothing abnormal would be noticed by the crew. Within a most intense gravitational field the same would happen—time would simply stand still. Once again that "ever-flowing stream" would have become frozen.

Where could such an intense gravitational field be found? Almost certainly close to the boundaries of that strange phenomenon popularly known as a black hole. From one of these strange objects nothing, not even light, can emerge. Since we will be considering black holes in some detail in the following chapter, I'll not probe this aspect of the strange universe too deeply at this juncture save to remark that the ramifications with respect to time and space are truly enormous. It assuredly would be a very courageous (or foolhardy!) bunch of astronauts who would be so bold as knowingly to permit their spacecraft to drift into a black hole. It may seem very much like an elaborate and somewhat spectacular form of suicide, but once we begin to investigate the probable form of a black hole's structure and interior, we may become inclined to change our minds. Unfortunately, the nearest black holes, so far as we are aware at present, lie at vast distances from the Solar System. To utilize these as a means of time, interstellar, or intergalactic travel would first involve our reaching one—a clear case for employing travel at relativistic velocities as an essential preliminary.

It is not unreasonable at this point to inquire whether there could conceivably be any other way in which time somehow could be stretched or distorted to our advantage, obviating the need to make a two-way trip of several light-years. Theoretically, vibration or oscillation may be shown to have an analogous effect on the flow of time, but unfortunately this would necessitate frequencies so great that material objects would almost certainly be rent asunder by the tremendous strain.

TUNNELS IN TIME AND SPACE

A star-ship dives or, more correctly, is sucked into a black hole. Is this the end of that vessel and its occupants? Not necessarily, for reasons we will be seeing, but it would almost certainly be the end of all normal existence for those aboard. As the craft emerged from the other side of the hole (we will see later what form that is likely to take) its occupants could have the bizarre experience of finding themselves in a remote part of the universe—or in another time, past or present.

First, however, it is essential to discuss the nature of black holes. Although covered at some length in one of my earlier books, *Interstellar Travel: Past, Present, and Future,* an updated version is offered here—updated, because in the years between that book and this, the amount of literature on black holes and their effects has increased considerably.

The existence of black holes was first postulated seriously in 1939, but as long ago as 1798 the renowned mathematician and astronomer Pierre Simon de Laplace (1749-1827), had suggested their presence, claiming the likelihood of stars having such tremendous dimensions and gravitational fields that not even rays of light could escape their clutches. As a consequence they would be rendered invisible. Not surprisingly, in 1798 little if any credence was accorded to such a remarkable idea. Now, close on two centuries later, the existence of black holes is largely accepted by cosmologists. We can hardly do better than quote the words of Dr. Kip Thorne of Cal Tech, who describes them in the following graphic and eloquent terms:

> Of all the conceptions of the human minds from unicorns to the hydrogen bomb, perhaps the most fantastic is a black hole—a hole in space with a definite edge over which anything can fall and nothing escape. It is a hole with a gravitational field so strong that even light is caught and held in its grip—a hole that curves space and warps time. Like the unicorn and the gargoyle, the black hole seems much more at home in science fiction or in ancient myth than in the real universe. Nevertheless the laws of modern physics virtually demand that black holes exist. In our galaxy alone there may be millions of them.

These words were written in 1974. It was only in 1970 that what may be proof of the existence of a black hole emerged. During that year data from a satellite revealed that a strong X-ray source in the constellation Cygnus, known as Cygnus X-1, might well indicate the presence of a black hole/supergiant star binary system. Its distance from us—6,000 light-years. It is generally believed that the hole orbits a giant star known as HDE 226868. To be precise, they orbit each other. Just why X-ray emission should be indicative of a black hole is something we will come to shortly. The questions to which we must first address ourselves are how and why black holes originate.

As any star ages (and that includes our Sun) it consumes so much of its hydrogen in nuclear reactions that a highly critical and "dangerous" state is eventually reached. This results in the star's swelling out into a great bloated red body known to astronomers as a "red giant" (ßetelgeuse in Orion and Antares in Scorpio are classic examples). This is the fate that will eventually overtake our own familiar benevolent Sun—only by then the Sun will have become far from benevolent. Since this will not occur for another 5,000 million years, there is no need to press the panic button. As steadily more and more of a star's nuclear fuel is consumed, including helium and certain of the heavier elements in addition to hydrogen, the red giant's expansion process can no longer be maintained and contraction begins. With a star of moderate mass, such as the Sun, contraction does not merely continue until the star reverts to its former proportions. The process continues inexorably until it eventually possesses a diameter comparable to that of Earth (8,000 miles). When a star has attained this stage it is termed a "white dwarf." Obviously, with such an enormous amount of matter contained within a body of relatively minute proportions, the density must have increased enormously. The density of a white dwarf star is so tremendous that it is virtually impossible to comprehend. The matter in such a star is said to be in a degenerate state. Observations indicate densities of the order of 100,000 million kilograms per cubic meter. A tennis ball filled with its material would have the mass of a naval cruiser!

The expansion-contraction process from normal star to red giant and thence to white dwarf involves a star of less than 1.4 solar masses (Chandrasekhar's limit) prior to gravitational collapse. However, incredible though it may seem, a white dwarf need not necessarily represent the ultimate stage in stellar evolution; a cataclysmic event known as a supernova may result if the star is sufficiently massive. Supernovas are of two types, I and II. The cause of Type I is rather uncertain but may be attributable to a white dwarf that accretes sufficient mass from a close companion star to exceed Chandrasekhar's limit. Alternatively, it could be the core of a supergiant star which, having lost its outer envelope, suffers an explosion similar to a Type II.

Type II supernovas occur only in the spiral arms of galaxies. A pre-supernova is almost certainly a supergiant in excess of 8 solar masses, putting it well beyond the 1.4 solar masses that lead to the eventual formation of a conventional white dwarf.

So tremendous is the explosion of a supernova that up to 90 percent of the star's mass may be ejected into space and for several days it may outshine an entire galaxy! What remains thereafter is a collapsed stellar core at the center of a rapidly expanding shell of gas. A prime example of this sort of object is the well-known Crab Nebula, 4,500 light-years from us in the constellation Taurus. Another is the Vela supernova remnant, 1,500 light-years distant at its center, but only 300 light-years away at the nearest edge of its nebula. Since a stellar core is far too compressed and small to constitute a white dwarf, it finds equilibrium instead as a "neutron" star, a body so minute that it may possess a diameter of only about *12 miles.* Needless to say, the density of such an object reaches colossal proportions—about a *hundred million* times that of a white dwarf. Black holes most likely result also from very massive stars that have exploded as supernovas, leaving a core in excess of 3 solar masses. Such a core must experience *total* gravitational collapse since it exceeds the stable limit for both white dwarfs and neutron stars. In the case of the Cygnus X-1 system, a 20-solar-mass supergiant is believed to be accompanied by an invisible companion having a mass some 6 to 10 times that of the Sun, i.e., a black hole. To imagine density greater than that of a neutron star is well nigh impossible. In these circumstances standard gravitational theory must be regarded as inadequate. In its place relativity theory must be invoked.

A black hole must therefore be regarded as a region of space into which a massive "dying" star has collapsed, and in order to decline to this state a main-sequence star "originally" of 10 to 30 solar masses must be involved. The principal feature of a black hole is thought to be a spherical surface with a radius proportional to the mass of the hole (see Appendix 2). This surface is known as the "absolute event horizon." Within lie the peculiar remains of the crushed star to which the hole owes its existence. As we will see, the remains of the star are not in the literal sense remains at all but something infinitely more strange. Due to the peculiar phenomenon of infinite time, itself the consequence of unimaginable gravity, light cannot escape. Even as the star is collapsing light rays are being seriously affected by the "embryo" black hole (Figure 14).

The effect of the intense and increasing gravitational field within the event horizon is that light is "clawed" back unless its beams lie within the "zone of escape." The more closely to the event horizon that light rays are emitted, the greater is the tendency for them to be pulled back toward the center of the black hole. Some, though failing to escape completely, are not pulled into the hole. Such rays will be unable to radiate into space and will instead constitute a concentric sphere of "frozen light" (a photon shell).

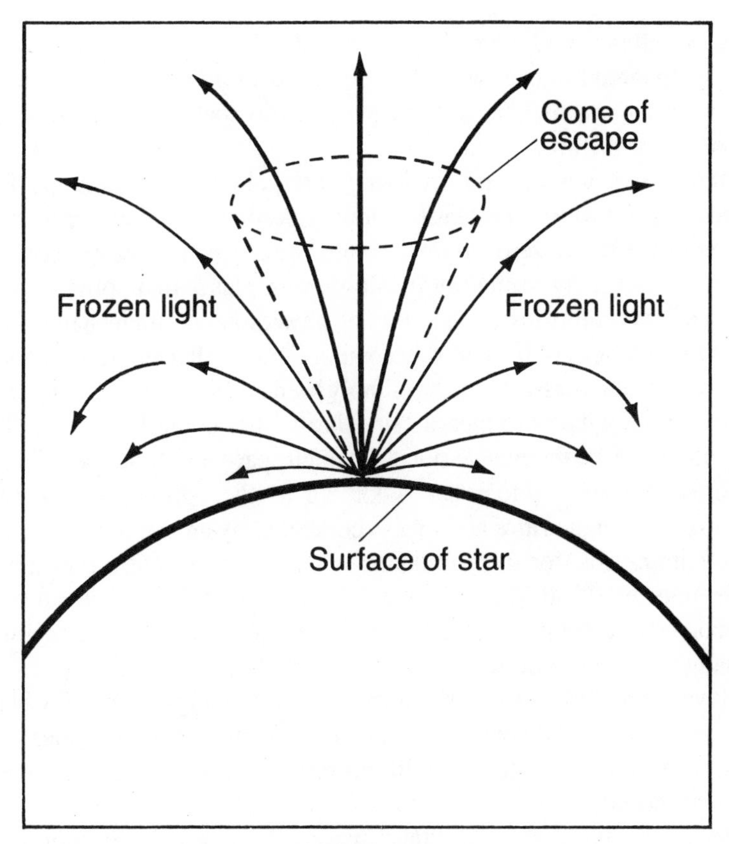

Figure 14

Light rays leaving the collapsing star/embryo black hole in paths within *the "cone of escape" continue to radiate. Those outside either form a concentric sphere of "frozen light" or are so bent as to return to the star.*

As the star continues to collapse, space-time becomes warped ever more severely. The diameter of the escape cone, as might be expected, shrinks progressively; as it does so, increasingly fewer light rays are able to emerge. In the final stages of collapse only those at or very close to the vertical can escape. Ultimately the process reaches a critical point at which the cone of escape simply ceases to exist. From then on it makes not the slightest difference at which angle a ray of light is emitted, for the now-prevailing circumstances will *never* permit its escape, nor for that matter will anything else. All has become permanently locked within an intense gravitational prison.

Across this unique and terrifying threshold the gravitational field increases still further in intensity. Gravity becomes infinite, crushing the matter of the star to zero volume. To use terms like "infinite gravity" and "zero volume" may seem absurd, but we must realize that nothing within a black hole conforms to the laws of physics as we understand them. Pressure and density have also become infinite so that the star involved, once a great and massive Sun, has simply been crushed into "nothingness." At this stage what is termed a "singularity" has come into being. This is defined as a mathematical point at which space and time are infinitely distorted. According to what is known as the "principle of cosmic censorship," singularities are always concealed by an event horizon, and so it is impossible for them to communicate their existence to observers in the universe. "Naked singularities"—i.e., singularities *without* an event horizon—have been suggested, but the existence of such phenomena remains a subject of debate.

Having briefly described the formation and configuration of a black hole, we must now consider its possible employment as a means of achieving a form of time travel. In this respect black holes could be very relevant indeed.

We will assume that an interstellar spacecraft has reached the immediate environs of a black hole. As already mentioned, clocks aboard the vessel will run slow. The more closely the ship approaches the black hole and its increasingly intense gravitational field, the more slowly will its clocks run. Time is simply slowing down. When the event horizon is reached, they stop—but only so far as we on Earth are concerned. If in some manner we could observe this strange drama unfolding, it would *appear* that the passage of the spacecraft was taking an increasingly long time. Indeed it would (to us) take an *infinite* time to cross the event horizon. Put more succinctly, the spacecraft would appear to be stuck there forever! Not that its occupants would notice anything untoward. Their clocks would appear to them to be functioning perfectly normally, and their own general, physical, and metabolic functions, having slowed down in phase, would appear equally normal.

By use of the intensely strong gravitational field surrounding a black hole, time dilation might also be achieved. This would involve plunging the spacecraft in and out of the field—the more passes, the greater the cumulative advance into the future. Note that this does not involve entering the black hole, although the risk of inadvertently doing so would undoubtedly be very real.

This is not intended to suggest that the occupants of the spacecraft would not be in mortal peril of a totally unprecedented kind were they to enter the black hole, for they along with their ship risk being crushed into nothingness unless certain essential conditions are met.

This brings us back to the fundamental nature and structure of a black hole at the center of which lies the singularity. To collide with this (if "collide" is the correct term in the circumstances) would undoubtedly spell total extinction, both rapid and spectacular, for the spacecraft and its occupants. It is believed, however, that the certainty of being swept into the singularity arises only if a black hole fulfills two essential conditions: it is nonspherical, and it does not rotate.

For a number of sound reasons (which involve some rather involved mathematics to explain), it is highly improbable that a black hole could be other than spherical. So far as rotation is an issue, it must be stated that a nonrotating entity in the universe would be utterly out of step with every other; galaxies, stars, planets, moons, and asteroids *all* rotate. There is no valid reason to believe that a black hole, being the "remains" (albeit greatly altered) of a star, would behave differently. Rotation in the universe is apparently fundamental. This being so, a spacecraft, in theory at least, should be capable of traversing a black hole without encountering the singularity. Just how this would be accomplished will be explained later.

It is important at this point to address the most fundamental question of all: Do black holes *really* exist or must they be regarded as speculative, highly theoretical concepts? There would be little point in considering them as agencies for time travel if they did not exist.

Black holes are believed to be prolific sources both of X and gamma rays; indeed X-ray sources have come to be regarded as the hallmark of black holes, though this is not to suggest that black holes alone are responsible for X-ray emission in the universe. But why should black holes emit X-rays? Let us first consider the binary system Cygnus X-1, which includes what is very likely a black hole, and its blue supergiant companion star, HDE226868. The intense gravitational field of the black hole has the effect of drawing a "tide" from the surface of this star. This material streams toward the hole (it is believed), forms an accretion disc around it, and then spirals inward. As it descends ever farther into the hole it becomes increasingly hot. Finally, due to a combination of compression and friction, X-rays are radiated copiously.

Other perils apart, X-ray emission of such intensity would represent a considerable hazard to any space travelers. We can only trust that efficient and practical X-ray shielding will have been incorporated in spacecraft traversing these strange holes in space.

Since X-ray sources are not necessarily indicative of the presence of a black hole (neutron stars, white dwarfs, and binary and multiple star systems are also capable of radiating in this region of the electromagnetic spectrum), what other observational techniques qualify? A search of the heavens for something which by its very nature is invisible can hardly be a straightforward affair. No matter how powerful or sophisticated the telescope, it will not enable an observer to "see" a black hole. A blind "observer" would be no worse off.

Probably the most satisfactory alternative approach would involve the detection of a black hole by reason of the powerful gravitational effects it would exert on other stellar objects in the vicinity. Nevertheless, the presence of copious X-ray emission from a particular point in space still provides the most significant indication—especially if no visible body appears to lie at this point.

In all the binary X-ray sources pinpointed during the past two decades or so, the mass of the system has been of the order of 1.6 solar masses. This would indicate the presence not of a black hole but of a neutron star. Until quite recently the only fairly well proven exception was our old friend Cygnus X-1. In this system the compressed star is reckoned to be equivalent to 6 solar masses. The only doubt some astronomers have is that the visible star in Cygnus X-1 (HDE226868) might just be a very luminous supergiant. Because of the high luminosity, its supposedly invisible (black hole) companion might simply be rendered invisible by reason of the glare. This belief has not, however, gained wide acceptance.

Very recently another likely candidate for black hole status was discovered by John B. Hutchings and David Crampton of the Dominion Astrophysical Observatory along with Anne P. Cowley of the Arizona State University. They had been examining luminous X-ray sources in the Larger Magellanic Cloud. Incidentally, the Larger Magellanic Cloud (which can be viewed only from the southern hemisphere) is a small galaxy thought to be an adjunct of our own Milky Way, lying a mere 160,000 light-years from us (not a tremendous distance by galactic standards). It is about one-fourth the size of the Milky Way. Near to it lies the Lesser Magellanic Cloud, about one-sixth the size of the Milky Way and some 190,000 light-years distant.

This source of X-rays in the Larger Magellanic Cloud has been known for a number of years and is designated as LMC X-3. During this time one particular star has come under increasing suspicion as being the *visible* component of the X-ray source. Regrettably, early X-ray telescopes proved inadequate to provide sufficient data. Only when the orbiting Einstein Observatory pinpointed the position of the X-ray source to within a few seconds of arc was it considered justifiable to study this star more closely.

One of the first facts to be elicited was that the visible star was alternately moving toward and away from us (the Doppler effect). This would strongly suggest that it is orbiting an *unseen* object over a period of 1.7 days. The orbital characteristics measured by the three astronomers limit the range of the possible mass of the invisible companion. For any reasonable value of the visible star's mass, it seems that that of the unseen object must be greater. Moreover, the astronomers believe this mass must be greater than that of a neutron star. In their own words, "It appears that LMC X-3 is the most convincing case yet found for the elusive stellar black hole."

Now let's consider the time travel potential of black holes. In *Interstellar Travel* I dealt at some length with the possibilities of "tunnels" in space, routes that could effectively provide a very tangible shortcut from star to star even though the stars concerned were extremely remote from one another. Here we are going to focus on black holes as tunnels in *time,* a facet only briefly touched upon in that book. However, for the benefit of you who may not have read *Interstellar Travel,* I'll first recap and enlarge slightly on the theme of "space tunnels."

Due to gravity, the effect of even a normal star such as the Sun is to warp space in its vicinity. Where black holes are concerned this effect is infinitely more pronounced. Under such circumstances it often proves helpful to envisage a model. Try to imagine a great plain on which, scattered around at random, are a number of caldera-like depressions. The plain is in fact space, the calderas being stars and collapsing star/black hole systems. In passing across this great plain (i.e., journeying in interstellar space), we might by chance encounter one of these strange depressions and fall into it. If it is one analogous to our Sun, a spacecraft would then spiral into the star or more likely orbit it, as do the planets. Should, however, the depression be due to a black hole the situation is radically different. Black holes, because of the tremendous intensity of their gravitational fields, cause very severe warping or curvature of space-time. Now we are confronted by a most fascinating possibility, for it could be that if space-time is sufficiently curved and warped, we might conceivably find a pathway into another part of the universe—or even as we'll see later, into an alternative universe.

A black hole is believed to develop by stages, so we must consider the possible nature of those stages. Most important, we want to know what the end result of the process might be, for this has tremendous implications

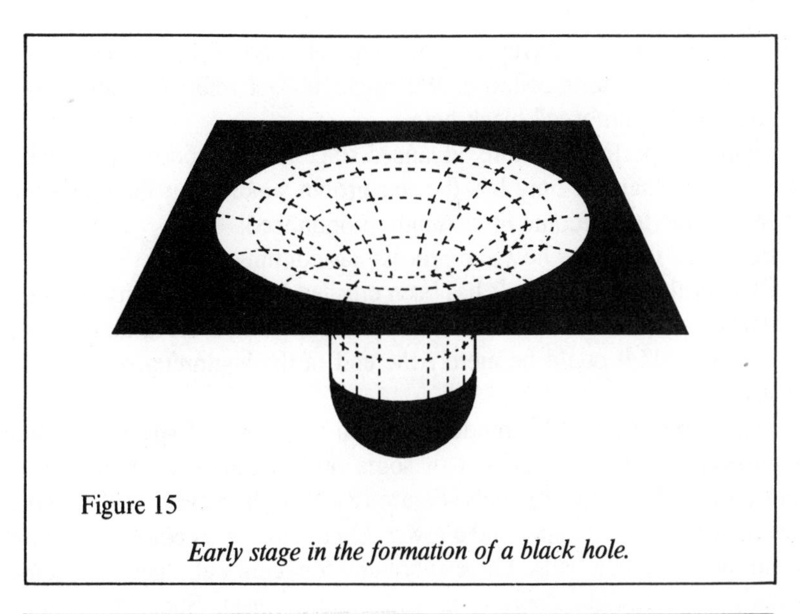

Figure 15

Early stage in the formation of a black hole.

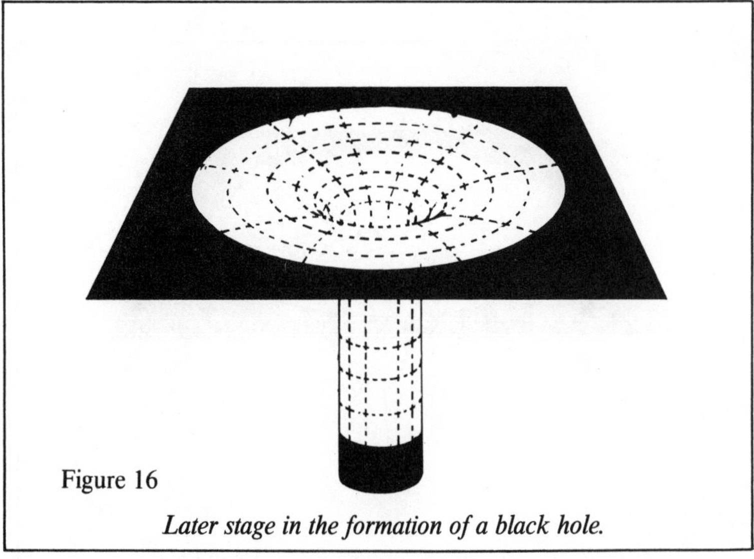

Figure 16

Later stage in the formation of a black hole.

for both interstellar travel *and* time travel. Figure 15 illustrates the beginnings of a star's collapse. We might in fact refer to some of the depressions as "embryo" black holes.

Figure 16 portrays a more advanced stage in the process of collapse. It is evident that not only has the star grown smaller but the curvature of space-time has become increasingly pronounced.

Figure 17 illustrates how the process has continued.

By now the star is gone. It has been crushed into a singularity. All that remains is an even greater degree of curvature—and a black hole. Is this really the end? It could be merely the end of the beginning, as we shall shortly see.

Earlier we envisaged a model in which the fabric of space-time was simulated by a sheet of rubber or some other flexible material. Let us return to this for a few moments (Figure 18). Now, however, we'll imagine *two* such sheets—an upper and a lower. On the upper, as before, we place a number of tennis balls. These represent conventional, "healthy" stars. Each causes a slight depression in the upper sheet. Such depressions

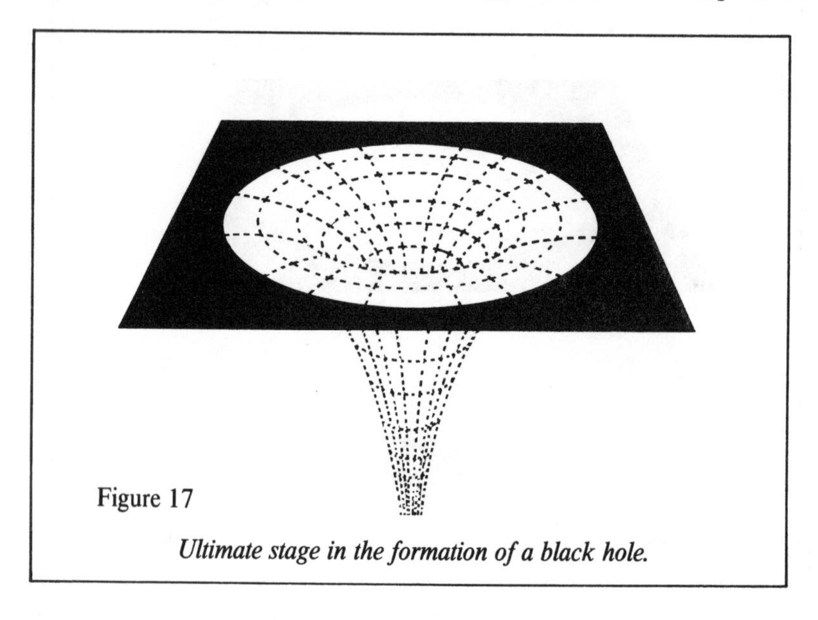

Figure 17

Ultimate stage in the formation of a black hole.

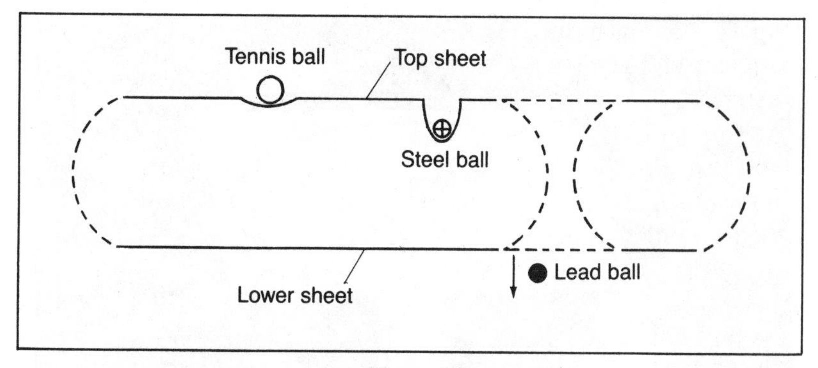

Figure 18

represent the curving or warping of space due to gravitational effects of a fairly moderate nature, e.g., that caused by the Sun. Suppose now we take two very much heavier balls, one made of steel and another of lead, and note the effects of each on the sheet. These are fairly predictable. The steel ball will create a very deep depression in the sheet. This we can liken to a black hole in the making (Figures 15 and 16) where the gravitational field is very intense and the space warping much more marked. The lead ball, however, sinks so deep and stretches the material to such an extent that a hole is ripped in the sheet. The lead ball now disappears from sight, and the plastic sheet, though it now contains a hole, springs back to its original flat state (Figure 18). What now of the lead ball? It continues on its way and rends the lower sheet also. How can we express this in cosmological terms? The top sheet represents the normal space-time continuum with which we are all familiar. The bottom sheet is also, for our purposes, part of the space-time continuum, though very remote. The point is this: the lead ball has disappeared from one region of the universe, and rending the lower sheet it has now appeared in another. The essential feature to grasp is that to get where it is now, a distant region of the universe, it has not been required to traverse the long, interminable light-years of space in the conventional way. It has taken a shortcut. And this shortcut we term a form of hyperspace.

In Figure 17 a fully-fledged black hole was portrayed as tapering to

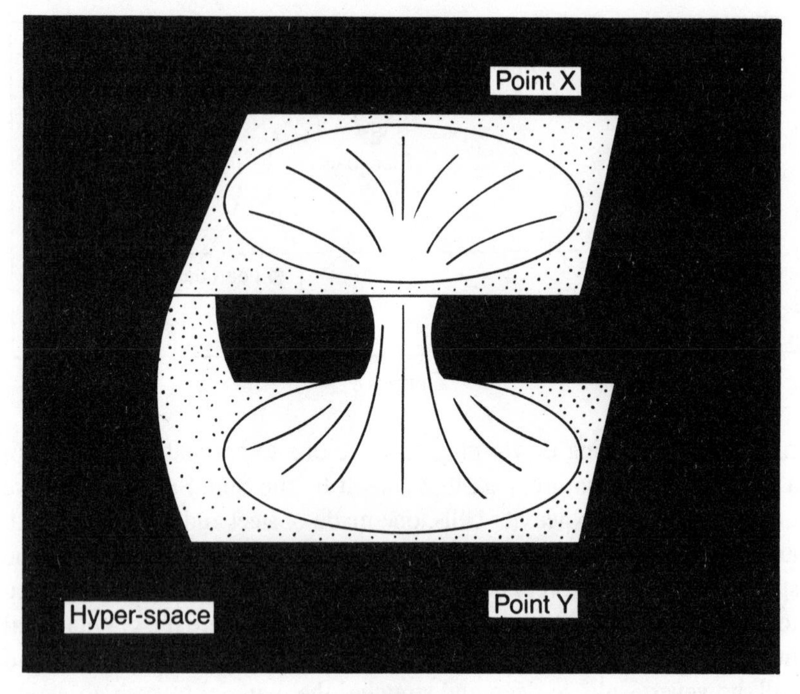

Figure 19

Representation of an Einstein-Rosen bridge whereby a "gravitational" tunnel connects two remote regions of the galaxy.

a point at which lay the totally crushed star or singularity, and we posed the question whether or not this was *really* the final state of affairs. If we take two models of the black hole as it appears in Figure 17 and place them so that one is effectively the mirror image of the other, something very interesting results, as Figure 19 clearly illustrates. That something is akin to the shortcut or tunnel in space mentioned with respect to Figure 18. This is known as an Einstein-Rosen Bridge.

In 1935, Albert Einstein and his colleague Nathan Rosen published a paper predicting the interconnection of the universe by *timeless* tunnels, "shortcuts to infinity," as a skeptical contemporary remarked. More than half a century ago such a concept was hard to accept and even harder to visualize. Nevertheless, it endured and is now looked upon with considerable if not universal favor.

To clarify this somewhat we will employ a further analogy. It is fairly certain that most if not all of us must have observed that when allowing bath water to run to waste a small but quite definite whirlpool forms just above the outlet. Closer observation will reveal that at the center of this there is *no* water. It is all swirling around the sides. If one very carefully inserts some long, extremely slim object such as a knitting needle precisely down the center of the vortex and then withdraws it, chances are it will emerge dry. Were it feasible to photograph the vortex, it would be apparent that the dry core narrows with depth until at the bottom it becomes closed off completely. Inserted too far, then, our needle would assuredly get wet. The vortex would in fact resemble to some extent the upper portion of an Einstein-Rosen bridge. Here the analogy ends, for an Einstein-Rosen bridge does not taper off at the center although it narrows considerably.

Now we come to what, I suppose, could be dubbed the 64 dollar question. If a black hole constitutes the upper section of an Einstein-Rosen bridge, how does the lower or mirror-image portion manifest itself if it opens up in a remote part of the galaxy? A black hole, as we have seen, sucks up everything in its immediate vicinity and will not permit even light rays to escape. Since an Einstein-Rosen bridge is regarded as a shortcut into deep space, its lower or "exit" portion can hardly be a *black* hole. This emergent end has been termed a *white* hole, from which light and matter are reputed to gush. Just as nothing can leave a black hole, nothing can *enter* a white hole. A journey through a black hole/white hole system (an Einstein-Rosen bridge) must therefore be regarded as a strictly one-way trip unless—which is highly improbable—another black hole lies conveniently near the white hole, thereby enabling a space-time voyager to return and emerge from a white hole not too far from where he started. This seems to stretch credulity much too far.

The next question is obvious. Since black holes seem to exist, where are the white holes? Certain highly energetic objects such as quasars have been proposed as possible examples. These are compact extra-galactic objects resembling points of light, though emitting more energy than a *hundred galaxies.* Since their light-producing regions are estimated as no

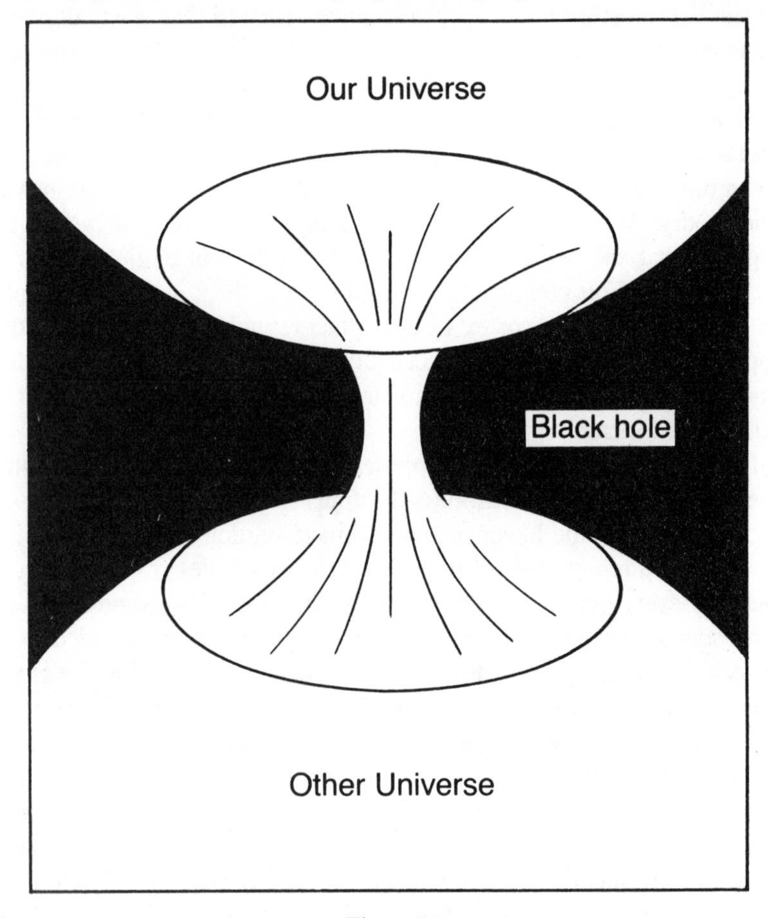

Figure 20
How an Einstein-Rosen bridge might constitute a passage between our universe and a parallel one.

more than a light-year across, it is extremely difficult to come up with a satisfactory explanation for such concentrated energy emission. It is almost as if a white hole served as the emergent end for a linkage of *several* black holes. Bear in mind that a single light-year is the equivalent of 6 million million miles. Could there be something in this line of speculation? These are still very early days in this field of cosmological research, and we should retain open minds. For all we know, large white holes could be responsible for the strange emissions currently being observed in the nuclei of galaxies. White holes should emit light continually, but at present it is difficult to predict just how bright they could be. Some could be very bright and others so dim as to be well nigh invisible. Obviously, the parameter of distance will enter into the picture, but it could be that less matter is being ejected from some white holes than from others.

Figure 19 portrayed the manner in which an Einstein-Rosen bridge could constitute a cosmic "by-pass" within our own galaxy. Some cosmologists believe that this concept could have much wider ramifications—that Einstein-Rosen bridges might even lead to another universe (Figure 20). The possibilities of parallel universes vis-à-vis time travel will be discussed later. For the moment we will confine ourselves to the implications of such links within our own universe. At first both Einstein and Rosen were reluctant to accept that such a bizarre entity as a "bridge" could possess real physical meaning. They were, however, thinking specifically in terms of a *nonrotating* black hole in which any object crossing the event horizon would be crushed out of existence by the singularity. When the much more likely concept of a *rotating* black hole was envisaged, the possibilities were greatly enhanced. But just why should the factor of rotation bring this about?

A nonrotating black hole, as we said earlier, is a very unlikely entity because everything else in the universe is in rotation. Before discussing the much more likely *rotating* black hole, it serves a purpose now to observe the hypothetical journeys of three spacecraft as shown in Figure 21. The black hole depicted here is of the *nonrotating* sort.

Ship A avoids the black hole completely, follows an absolutely normal

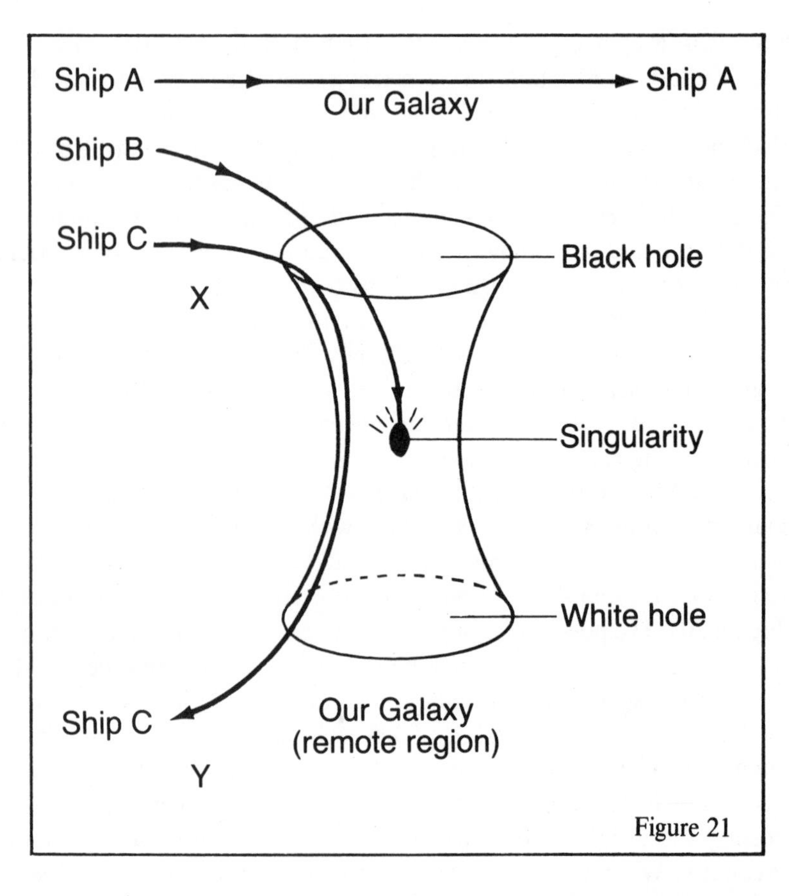

Figure 21

course, and proceeds on its uneventful way. Ship B is considerably less fortunate, and that assuredly is the last we are ever going to see of it and its foolhardy crew, for, having crossed the event horizon, it is quickly crushed out of existence by the singularity into which it is drawn. An entirely different set of circumstances prevails in the case of ship C, which somehow succeeds in avoiding the singularity after crossing the event horizon. By so doing, the ship and its crew should, with reasonably good fortune, emerge in another part of the galaxy entirely. It will have proceeded from point X to point Y in the conventional space-time continuum *without having traversed the intervening distance.*

Remember that the above scenario relates to a nonrotating black hole—something now regarded as highly improbable. Once more, as with everything else in the universe, the star that collapsed to create the black hole must also have been in rotation. Indeed as the process of collapse gathered momentum so also would the rate of rotation. But why should a rotating black hole be so different? The reason, expressed in its simplest terms, is that rotation has the effect of so distorting or warping space-time that it could effectively create *safe* avenues into other regions of space and of time.

In a rotating black hole there are not one but *two* event horizons, an

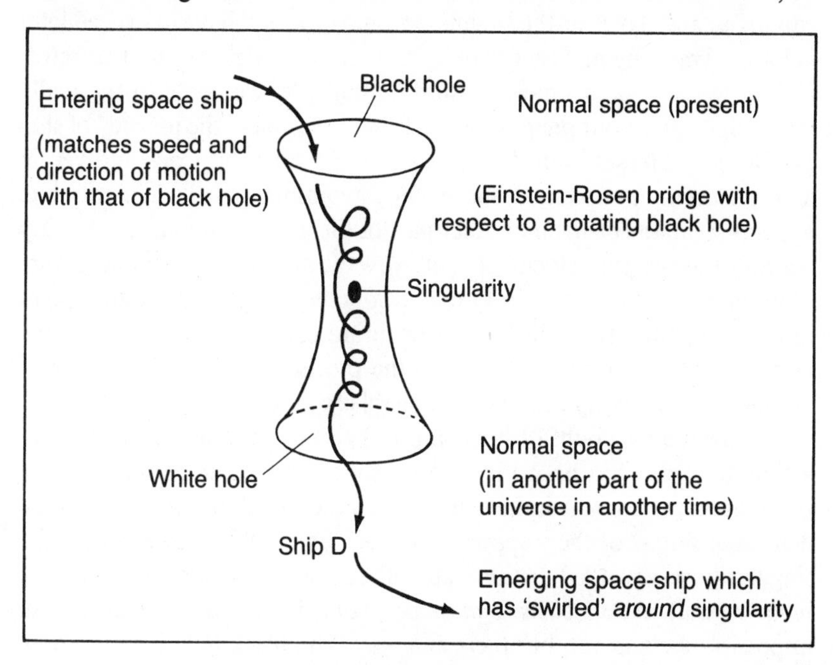

Figure 22
The path of ship D.

outer and an inner. Rotation has effectively split the event horizon into two parts. Let us see now what happens to our hypothetical spacecraft (Figure 22). In this instance the fleet consists of four ships—A, B, C, and D. As previously, ship A merely travels along in normal space-time on a safe journey. Ship B and its crew cross the inner event horizon and, as before, hurtle into the singularity—end of story! Ship C is not very lucky either. In Figure 21 it was made to appear that in a nonrotating black hole it should have got away with it. On this occasion, ship C passes the inner event horizon but is able to reach another part of the universe only if it is able to exceed the velocity of light, which we accept as impossible. Why, it might be asked, is the velocity of light involved? Because the event horizon is by definition a boundary from which light cannot escape. To clear the singularity, ship C would have to exceed light velocity. Traveling no faster than light, its crew's endeavors are frustrated, and so we will see no more of ship C. But now we come to the really interesting part of our imaginary space and time epic—the journey of ship D. This ship crosses both the outer and inner event horizons and yet, in a way soon to be explained, successfully avoids the lethal singularity and appears *simultaneously* in another part of the galaxy. Moreover, the ship has *not* exceeded the velocity of light. Now the most mind-boggling feature of all must be revealed: ship D may *also* have emerged at another point in *time!* Whether this will be past or future compared with the starting point in space-time may depend on the precise route the ship followed around the singularity and through the black hole.

Although an account of how ship D became both an interstellar and a time transport has been given, no mention has yet been made of the precise manner by which the ship avoided the singularity that caused the demise of ship B or the trapping of ship C. Figure 22 illustrates how the singularity was avoided. In the center of the diagram we see the singularity lying in wait. At all costs it *must* be avoided. This may seem like an impossible assignment, but there *is* a way. The black hole is in rotation, so what the crew of a spacecraft *must* do is to match the speed and direction of their ship with that of the hole's rotational velocity. By so doing the ship would swirl (like the bath water) *around* the singularity and thereby avoid it. The ship could then emerge in another part of the universe, although its crew would have no choice as to which part. It could go only where this strange space-time tunnel led.

It should not be supposed that entering a black hole under these conditions is going to prove anything but a very considerable feat of astronautical skill. For example, a black hole equivalent to 10 solar masses would have a diameter of barely 37 miles and a navigable diameter of only 600 meters (see Appendix 2). Readers may wonder how such precise values are calculated. These follow from an equation derived as long ago as 1916 by the German physicist Karl Schwarzschild. The result generally is referred to as the Schwarzschild radius. This may be defined as the radius of the event horizon—a critical value that must be exceeded by a body of light from its surface is to reach an external observer. For a body of mass M, the Schwarzschild radius R_s is given by the equation

$$R_s = \frac{2GM}{c^2}$$

where G is the gravitational constant and c the velocity of light. Should a star collapse so that its radius becomes less than the critical value (R_s), the escape velocity becomes equal to the velocity of light and the star becomes a black hole. The radius R_s, as the equation shows, is proportional to the star's mass. In the case of the Sun it would be slightly under 2 miles (3 kilometers), allowing a navigable aperture of a mere 30 meters! On account of its relatively low mass we know that the Sun could never degenerate into a black hole, but if it could, this would be all the room for maneuver that a spacecraft would have.

Another aspect relating to black holes and their relevance to space-time travel must be mentioned. This concerns the period during which Einstein-Rosen bridges are able to endure. Do they last indefinitely or are they more likely to possess only limited duration? Present cosmological thinking tends to the belief that they could endure for much longer than the present expansion-contraction cycle of the universe but of course could not if the universe contracted into a new embryo primeval "fireball." This will be gone into more fully in Chapter X. Obviously, if the universe expands indefinitely, Einstein-Rosen bridges would have as much time as they required.

It is thought that initially an Einstein-Rosen bridge is large (depending of course on the collapsing star's original mass) but in due course begins to contract fairly rapidly by cosmological standards. Eventually contraction is complete with the tunnel blocked or "pinched off" at its narrowest point. There is then no way by which a spacecraft could make it through and out of the white hole. It would simply be trapped, unable to go either forward or back—a most unenviable situation. Moreover, as the width of the tunnel approaches the critical point, a ship in the region of maximum contraction would very likely be engulfed by the singularity.

In Chapter V we dealt at some length with the speculative possibilities of a fifth dimension, termed "hyperspace," which could lie beyond the "skin" of normal space-time. We saw how this might provide a form of shortcut between two widely separated parts of the universe, but we haven't yet examined how a spacecraft might enter this dimension. Clearly the most wonderful of shortcuts is of little use if there is no means of taking it. A black hole system might just prove the gateway. In this case the region just beyond the singularity could prove to be that peculiar no man's-land we term hyper-space, (i.e., the "throat" of the Einstein-Rosen bridge) with the white hole acting as the exit into that remote part of the universe. What would it be like within the hyperspace "throat"? Given the peculiar way in which the fabric of space-tme is distorted, hyperspace could turn out to be that timeless avenue whereby a spacecraft is enabled to pass from one point in conventional space-time to another instantaneously. In his excellent book *Black Holes,* Dr. John Taylor describes most eloquently what a space-time astronaut might experience in this strangest of strange regions: "The traveller into hyper-space has to leave all his usual notions of space and time behind him. He cannot ask if hyper-space is hot or cold, whether it is wide or narrow or whether it is shaped like a cube or a sphere. It has no past or future nor any dimension. It is a lace-work of worm holes, forming and disappearing, constantly in motion but never advancing or retreating. It is full of ceaseless activity, yet overall it is static and timeless."

Much more could be said regarding black holes, white holes, and Einstein-Rosen bridges. Indeed an entire volume could easily be devoted to this subject area alone. In this chapter the description has been deliberately oriented toward their function as possible "time machines" as well as a means of reaching parts of the galaxy that would otherwise remain forever closed to us by the sheer parameters of time and distance. It is therefore now appropriate to revert to our real brief—travel in time. This does not mean we are done with black holes. Their capacity in this respect will be greatly extrapolated, enabling us to see how travel into the past, into the future, and even into parallel universes might be achieved. Already in Chapter VII we examined a possible method of entering the future by means of time dilation. Travel into the future via a black hole represents a wholly different concept and whether or not we could return to Earth is much more debatable—which probably would take care of all the causality problems!

A word of warning however. In his recent excellent book *A Brief History of Time,* Stephen W. Hawking, professor of mathematics at Cambridge University and probably the most brilliant theoretical physicist since Einstein, states, "This [black hole] would offer great possibilities for travel in space and time, but unfortunately it seems that these solutions may all be highly unstable; the least disturbance, such as the presence of an astronaut, may change them so that the astronaut could not see the singularity until he hit it and his time came to an end. In other words, the singularity would always lie in his future and never in his past." Clearly much research will require to be done before we hurl ourselves voluntarily into a black hole. Professor Hawking ends in a humorous vein, saying "it would mean [if this could be] that no one's life would ever be safe: someone might go into the past and kill your father or mother before you were conceived!"

TIME, LIGHT, SPACE, AND MOTION

Before proceeding further it would be advisable to coordinate our thoughts regarding the relationship existing between time, light, velocity, space, and motion. Much has already been said about the space-time continuum and the peculiar effects on it of intense gravitational fields, but it is necessary to tie all these factors together more precisely. One good way to do this is to look at Figure 23, a diagram that places the various parameters in their proper perspective within the space-time continuum. I confess I did not conceive this visual model; indeed, it is the type of space-time representation that with slight variations, has been used by relativist cosmologists over a number of years. It is known as a Kruskal diagram, after its originator. Notice that it is based on two principal axes at right angles to one another, the axes represent motion through space *and* time respectively. Each of us is passing through both space and time.

We may be seated and seemingly at rest, but we must take into account not only the Earth's daily axial rotation but also its orbital motion around the Sun. Neither should we be oblivious of the fact that the Sun and all its retinue of planets and other bodies is moving steadily through space toward a point not far removed from the lovely summer star Vega in the constellation of the Lyre. We are indisputably in motion through space. It is also obvious that we are simultaneously in motion through time. This is not something we can see or feel; only the effects of its passing are made known to us by a sequence of events or by reference to a clock or watch.

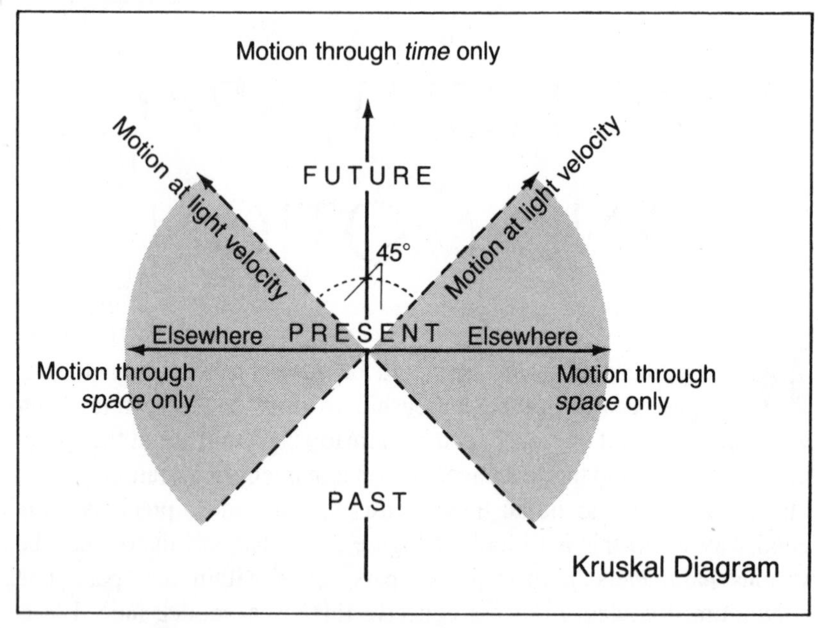

Kruskal diagram.

To use the diagram, however, we must learn to construe the facts in a somewhat different light.

The diagram portrays the relations of the space-time parameters to one another, but in a manner which may seem more than a little unorthodox. For example, you may at first be justifiably perplexed by the fact that motion through space and motion through time are shown at right angles to one another as if they were separate functions. This apparent anomaly we will return to later.

A Kruskal diagram is simply a convenient way by which relativists endeavor to portray on a two-dimensional sheet of paper something that has more than two dimensions. The dimension of space (the horizontal axis) has for practical purposes been rendered unidimensional. The same holds true for the dimension of time (the vertical axis). Thus, so far as the diagram is concerned, an object stationary in space moves only through time, and conversely an object stationary in time moves only through space. Now clearly such modes of "travel" are not feasible. What the diagram really sets out to do is to portray those journeys that *are* possible.

The diagonal dotted lines represent the velocity of light. This, as we have seen, is a limiting velocity. In other words, any journey at a velocity less than that of light is theoretically possible (it is best to say "theoretically" since at present it is impossible to approach anywhere near it). Thus we have "possible" and "impossible" journeys. The "impossible" type, it will be observed from the diagram, lies beyond the light velocity diagonal (as we would expect) and, since it is closer to the space axis, it is termed "space-like." This region is captioned "elsewhere." Because we cannot exceed the velocity of light, we just do not know what conditions "there" would be like. The "possible" type of journey is termed "time-like" since it lies closer to the time axis. (Later it will be demonstrated mathematically that if the velocity of light *could* be exceeded, time would run *in reverse,* carrying us into the past.)

It is necessary now to return to the journeys of the hypothetical spacecraft outlined in the previous chapter. Ship A, you'll recall, kept clear of the event horizon on its journey and so came to no harm—or at least no harm from the black hole. This represents a "time-like" possible journey since it was unable to exceed the velocity of light. Put another way, the journey of ship A is much nearer being parallel to the time axis. It is passing through space-time in a normal way. Ship B crosses the event horizon and is engulfed by the singularity. Its path on the Kruskal diagram, though not as close to the parallel of the time axis as that of ship A, is nevertheless much more nearly parallel to it than it is to the space axis. This ship too has not exceeded the velocity of light, and though the journey ends catastrophically, it must nevertheless fall into the category of "possible." Now we come to the journey of ship C, which though it avoided the singularity was unable to bring its journey to a successful conclusion since it too did not exceed the velocity of light. This was, therefore, another "time-like" possible journey.

At this point you may detect a seeming discrepancy in the argument. In the previous chapter this was the reason given for ship C's failure to complete the journey. But surely ship C, having crossed the event horizon of the black hole, should be able to escape. Its failure to leave the black hole because of its inability to exceed the velocity of light is certainly applicable were it desirous of recrossing the event horizon (nothing can emerge from a black hole). But this should not, it seems, inhibit its ability to leave via the white hole from which matter cannot enter and only leave. However, we must stress that within a black hole gravity is so strong around the singularity (where the normal fabric of space-time is crushed) that ship C could not pass this region of the black hole/Einstein-Rosen white hole system en route to the white hole exit *unless* it could exceed the velocity of light. This would represent a "space-like" journey and consequently would be impossible. Surely this must mean that no ship could use such a space-time "tunnel" since velocities in excess of that of light cannot be attained.

So far, however, we have been thinking purely in terms of a *nonrotating* black hole, which we accept as unrealistic in view of the fact that all other bodies within the universe are in rotation. When we come to ship D we will recall that it was enabled to pass through this system and emerge in another region of space-time because its crew matched the ship's velocity and direction of rotation to that of the black hole, thereby avoiding and passing the singularity. This rotation has the effect of distorting space-time in such a manner that an avenue of escape via the white hole to other regions of space-time is opened up. Thus ship D in this "avenue" does not require a velocity greater than that of light to pass through and reach these regions. It is swirled around and spewed out by the system from a region of crushed space-time fabric. Since its velocity does not exceed that of light, its journey is "time-like" and therefore possible. As already explained, the diagonal dotted lines on the Kruskal diagram represent motion at the velocity of light. This (tachyons apart) is a limiting velocity so far as anything having mass is concerned—and spacecraft have a lot of that.

Since on the diagram we are at the point marked "present," it is easy to see that all events that ever influenced the present must belong to that area in the diagram labeled "past." In like manner any actions taking place at "present" will influence the area marked "future." Because of the limiting velocity of light, the portions of the diagram *beyond* that velocity (between the rays of the 90 degree angle formed by the dotted lines on the right, and those on the left) represent "regions" of the universe which, though they may have some obscure physical existence, cannot possibly be influenced by any actions we take in the present, for we are unable to reach them. For this reason they are captioned "elsewhere." In the universe as we know it, we can recognize only past, present, and future.

All journeys are thus possible only in the remaining area marked "future," i.e., journeys made at less than the velocity of light. But what, you may be tempted to ask, about the remaining area labeled "past"? Surely this must also come into the category of "possible"? This is a reasonable and valid point and one with which we must shortly try to come to grips. For the moment let's accept that such journeys as can be made are restricted to the area marked "future." The scope of travel is thus limited. We can move forward into the "future" region but *only* at velocities less than that of light. Since we are presently unable to approach velocities considerably less than this, our freedom is restricted to an even smaller portion of the "future" region. If, in time, as it no doubt will, terrestrial technology enables us to attain velocities much closer to that of light, a much greater proportion of the "future" region will be open to us. This, of course, follows from what has already been said regarding space-travel at relativistic velocities so it really should not surprise us now. Remember, however, that this applied only to *normal* space-time. Within a black hole/Einstein-Rosen bridge system, conditions will be vastly different. Here travel into the future by a totally different technique could be feasible. In a sense this might be regarded as much less advantageous since we would almost certainly be unable to return to Earth to ascertain how this jump into the future had affected the world left behind. Time-dilation relativistic travel into the future, on the other hand, merely requires us to retrace our steps in space.

What actually has happened to ship D since it crossed the two event horizons? The ship and its occupants have entered a region where the normal laws of physics simply do not apply. It is one in which distance is almost totally negated, where all distance from the light-year to the millimeter becomes infinitesimally small, certainly too small to be measured by any instrument or technique so far devised by us.

"All right," you're allowed to object, "but surely this is still only travel in space, though a very remarkable way of achieving it. How does time travel fit into the picture?" So let's now correlate the two. In the case of ship D, a journey of many light-years has been reduced virtually to zero. In other words, to all intents and purposes the journey has been rendered instantaneous and *not* as one made at a velocity faster than light. Thus there is *no* violation of Einstein's Special Theory of Relativity. This also follows from the gross distortion of space-time in a black hole *whereby backward motion in time corresponds exactly with forward motion in space.* Thus a black hole can apparently lead into the future or the past—or so at least it is postulated.

For long it was assumed that whatever the chances of travel into the future, time travel into the past was an unrealizable dream—one so full of causality paradoxes as to render it ridiculous, but once the possible potential of black hole/Einstein-Rosen bridges is considered, the picture changes. Obviously it cannot be likened to the type of time travel into the past envisaged by some writers of science fiction. The time machine of H. G. Wells epitomized this. A craft was created capable of traveling through time in either direction without the baleful necessity of hurtling through space at relativistic velocities or plunging into rotating black holes. Naturally, the contradictions and paradoxes arising from causality were ignored—for they were many!

So now, in the next chapter, we will take a good hard look at time travel into the past.

TRAVEL INTO THE PAST

In the famous *Rubaiyat of Omar Khayyam* there are four lines that must be familiar to many:

> *The Moving Finger writes; and having writ,*
> *Moves on: nor all thy Piety nor Wit*
> *Shall lure it back to conceal half a Line,*
> *Nor all thy Tears wash out a word of it.*

Less well known perhaps are the lines that follow:

> *And that inverted Bowl we call the Sky*
> *Whereunder crawling coop't we live and die,*
> *Lift not thy hands to it for help—for It*
> *Rolls impotently on as Thou or I.*

The legendary Persian poet of the twelfth century perceived time in a fashion analagous to that forward ever-rolling stream we have already discussed.

It is hardly surprising that many physicists and cosmologists have shed considerable doubt on the notion of journeying backward into time, largely because of the clear violation of the laws of causality. We will return to this aspect shortly. But first, as a preliminary, we must look more closely into the position vis-à-vis travel at relativistic velocities. In Chapter

VII we looked only at the prospects for travel into the future. Let us now see precisely what happens to the basic relativistic equation when we consider travel into the past. In fact it is rendered nonsensical whenever velocities equal to or in excess of that of light are assumed. This presumably is as good a way as any other of demonstrating that such velocities are impossible within the normal space-time continuum. The relevant formula is as before:

$$\frac{T_{ss}}{T_t} = \sqrt{1 - \frac{v^2}{c^2}}$$

where T_{ss} is the time taken by a relativistic spacecraft, T_t is the time interval as measured stationary on Earth, v is the velocity of the craft, and c, as usual, represents the velocity of light. If v becomes almost equal to c, then v^2/c^2 tends toward (but can never achieve) unity. Thus the right-hand side of the equation is always a positive quantity and must therefore possess a square root. We can term this x, and clearly it must be a very small quantity. Thus $T_{ss} = xT_t$, indicating that a period of time on a spacecraft traveling with a velocity close to that of light is very considerably less than if it were at rest on Earth. This is precisely what we discussed in Chapter VII—travel into the future by means of time dilation. If the velocity of light *were* attainable, then

$$\frac{T_{ss}}{T_t} = \sqrt{1 - \frac{v^2}{c^2}} = \sqrt{1 - \frac{c^2}{c^2}} = 0$$

From this it would appear that the time taken by the spacecraft, so far as those aboard are concerned, is zero. This can hardly be regarded as realistic since it would imply instantaneous travel. Should v be greater than the velocity of light c, then the right-hand side of the equation becomes negative. Unfortunately it is impossible for a negative quantity to possess a square root.

The position therefore is that as v steadily increases, T_{ss} becomes progressively less (though still positive), but if it reaches c, T_{ss} becomes zero and from here on the situation is rendered ludicrous. *Could* c be exceeded without this mathematical impasse occurring in the progression, T_{ss} would become less than zero (i.e., negative) and by implication time would be running in reverse, present back to past instead of present into future. Since it is impossible to disagree with the basic mathematics, it must be accepted that whereas space travel at relativistic velocities does appear to offer a reasonably valid means of travel into the future it can offer none with respect to travel into the past. The only way whereby we can achieve this, it seems, is to find ourselves a conveniently placed black hole coupled to an Einstein-Rosen bridge/white hole system. Apart from the very real dangers inherent in crossing the event horizon there remains the question of reaching a black hole, for, so far, none has been located within a "convenient" distance of the solar System.

It has become perfectly obvious therefore that the basic equation has been rendered invalid and is now nothing more than an intriguing mathematical paradox. Such a proposition applied to *conventional* space-time is clearly ridiculous. However, a proposition of this kind does assume a certain validity if a spacecraft enters a black hole system, presupposing, of course, that its occupants have taken the essential safeguards mentioned previously. If they have, the singularity should be safely avoided with the ship emerging not only in another part of space but, depending on circumstances, *in the past* relative to the region it has left. It would appear that the conditions prohibiting travel backward in time do not necessarily apply within a black hole/Einstein-Rosen bridge system. Nevertheless, the equation is not being violated, for at no time has the spacecraft either equaled or exceeded the velocity of light. The previous journey (between the Sun and another star) is one between two widely separated points in *normal* space-time—i.e., it is a thoroughly conventional space journey. The moment we think in terms of a black hole system the position becomes radically and dramatically changed.

In the previous chapter we saw that within a black hole distance as such might be totally negated. Here we have apparently another paradox, for were this so, surely a "bridge" as such could not exist. We simply have to accept that in our dealings with black holes we have become involved with parameters of a totally unorthodox kind. We must recall that this is *not* a journey from A to B through conventional space-time. Some cosmologists now adhere to the belief that *forward* motion in space through the hole coincides exactly with a *backward* movement in time. Black holes are simply peculiar—a total negation or reversal of time and distance as we presently understand them. But to time travelers they could prove very useful indeed.

Whereas travel into the future produces awkward paradoxes if return to the present can also be achieved, travel into the past results in the most inexplicable state of affairs and a total violation of causality. But if travel into the past can be achieved only by recourse to an Einstein-Rosen bridge, the paradoxes due to causality are going to arise only if the travelers can in some manner return to the region of space-time from which they departed—and this is highly unlikely. This would seem to render the exercise pointless. When the travelers emerge from a white hole in *another part* of the galaxy, Earth could by then be back in the Stone Age and the travelers blissfully unaware of the fact. But since their mission might much more likely have been "to boldly go" and explore the galaxy in a search for other worlds, the time factor would be irrelevant.

It has been suggested that some beings of vast intellect on planets of other stars might be so technologically advanced as to be capable of creating their own personal "tunnels" in space, enabling them to roam the galaxy (or beyond) at will, this talent complete with the capability of returning. If some day in a distant future epoch our civilization here on Earth were able to achieve this near miracle, some of our remote descendants are going to be faced with the problem of causality for it is one that just will not go away.

It has been claimed by some physicists that objections to time travel based on causality violations have no real foundation. Paradoxes, they say, are due to the limitations of the human mind and the manner in which it functions. This would be a convenient way around a very tricky problem, but we can hardly dismiss causality so easily. The noted physicist and cosmologist William Kaufman subscribes to the traditionalist view in his excellent book *The Cosmic Frontiers of General Relativity:*

> Causality simply affirms that effects occur *after* their causes. If a light bulb in a room suddenly lights up, it is reasonable to assume that someone flicked the switch a minute fraction of a second earlier. It is absurd to suppose that the light bulb could light up because someone ten years in the future flicks a switch. The idea that effects could occur before their causes is denied by the rational human mind.

Kaufmann's words express very eloquently the outlook of the vast majority of human beings. This is hardly surprising. It has, on the other hand, been suggested that this outlook is purely philosophical, based on a particular accustomed *belief* about the nature of the universe—something similar to the religious, agnostic, or atheistic beliefs most of us carry forward from our early years.

When we are confronted by black holes, Einstein-Rosen bridges, and time dilation, we are being asked to deal with phenomena that are totally out of accord with the universe we understand—or think we understand. We cannot visualize them—only accept them. Until someone improves upon or replaces Einstein's General Theory of Relativity—and at present this seems highly improbable—there is no alternative. If rotating singularities exist, and there is increasing reason to accept that they do, then the possibility of time travel, including travel into the past, has to be accepted, however fantastic it may seem.

We have seen that the basic relativistic equation

$$\frac{T_{ss}}{T_t} = \sqrt{1 - \frac{v^2}{c^2}}$$

becomes invalid whenever v becomes equal to c, and for this reason travel into the past via the conventional space-time continuum is impossible.

It might be of interest, however, were we able to overcome this barrier, say, by means of a "tachyon drive." Tachyons, as we saw earlier, are believed to exceed the speed of light, but the idea of a tachyon-propelled spacecraft remains fanciful due to the mass of the craft and the fact that tachyons when approaching the speed of light are slowing down, ceasing to exist when they attain it.

A. As the spacecraft accelerated toward light velocity, its occupants would age progressively more slowly (not that they would notice). This is time dilation as discussed in Chapter VII.
B. When light velocity had been attained, time, so far as the vessel's occupants are concerned, would cease. Without being aware of the fact they would theoretically be able to traverse the entire universe without aging further—a seeming second to them would have become eternity.
C. At velocities greater than c the ship and its occupants should move backward in time. Just how this would feel to them is hardly imaginable. They would, in fact, be able to return to Earth *before* they set out on their journey. Perhaps, the intrepid Miss Bright could complete her round-trip after all!

We now see causality in its starkest sense and appreciate, if we have not already, the appalling tangle that would ensue could we journey into the past other than by a black hole to another part of the galaxy. Miss Bright has effectively become *two* Miss Brights—the one who "started one day in a relative way" and the other "who returned on the previous night." But surely the *two* young ladies must be one and the same? Common sense tells us that two people cannot be one and the same. One has yet to make the trip and can therefore have no recollection of the journey, whereas the other has. They cannot be one and the same. Eventually this would lead to an infinite number of Miss Brights.

The girl was Bright, but not bright,
And she joined in next day on the flight,

So then two made the date,,
And then four and then eight,
And her boy friend was stricken with fright.

This is typical of the type of problem that arises when causality is invoked. Reason and rationality are at once reduced to chaos.

We have stressed more than once the fact that time and space cannot be regarded in isolation. We are dealing with space-time. Consequently a "timecraft" must also be a spacecraft. This, of course, is at total variance with the type of timecraft envisaged by H. G. Wells in his *Time Machine.* The following example portrays rather well what would happen could this fact be ignored. A hopeful time-machine inventor somehow succeeds, using a sealed capsule, in traveling six months back into the past. Observing that six months have elapsed according to his instruments, he impetuously opens the door of the device and is at once rendered very dead! In the six months he has gone into the past, Earth has *receded* six months in its orbit and now lies in space on the other side of the Sun. Presumably, had he gone six months into the future, a like fate would have befallen him. It could be suggested that had his journey been twelve months into past or future he could have emerged from the time capsule unscathed. Unfortunately, since the Sun and the Solar System are also in motion through space, this would-be solution too is invalid.

What then about the original Miss Bright? Since "she travelled in a relative way," she also would have to be an accomplished astronaut capable of guiding her ship safely back to Earth's position "on the previous night."

There seems no valid way that the integral components of time and space can be divorced from one another. So fundamental is space-time that it is simply absolute—something that just cannot be split in any way.

Causality, by reason of the paradoxes it creates, renders time travel into the past a very dubious business. It appears that it is *not* a feasible proposition so long as we think only in terms of *conventional* space-time. It might be argued, of course, that the traveler on reaching his temporal destination could suffer total and permanent amnesia regarding the "future" he had left. He would simply be back in a past year, and then his life would start a complete rerun—one in no way different from that he had experienced previously. Naturally he would revert to the earlier year and yet another rerun whenever he reached the year of his departure for the past. Would this not be the answer to the search for eternal life? It would, however, hardly constitute a practical method of time travel though it certainly seems to solve all the causality problems. Well, not quite all! We are still left with the Miss Bright syndrome, for would not our traveler meet up with him or herself? (Those of you who have seen the Michael J. Fox film *Back to the Future* have seen a depiction of a near-miss in this regard.) Journeys at relativistic velocities are another matter, but these, as we have seen, can lead only into the future. Here too causality rears its head. It may be that, despite our efforts, we cannot reshape the future, attempts to do so merely acting as a form of negative feedback leaving the predestined future unchanged—or, as we have seen already, ahead of us could lie a *selection* of futures. We'll explore that prospect more in the next chapter.

From what we have seen so far only a black hole/Einstein-Rosen bridge system is likely to make travel into the past possible. Unfortunately black holes lie, so far as we are presently aware, at very considerable distances from the Earth and the Solar System. Prospective time travelers into the past must therefore first reach one. Only relativistic space travel is going to get them there within a human lifetime, the accomplishment of which will first take them some way into the future relative to Earth. The idea of being able to create *artificial* black holes is attractive, but the level of technology that would be required makes it difficult to see this technique as anything other than a science-fiction concept necessitating the employment of titanic forces. We must also ponder whether the creation of a small artificial black hole would necessarily lead to the creation of an Einstein-Rosen bridge and white hole. It is impossible for us usefully to speculate on what might have been achieved by alien beings with technologies millennia ahead of our own. We might also justifiably wonder whether an artificial black hole created in space too close to Earth might not constitute a grave menace to our planet. It is reckoned that one created on the surface of Earth, however tiny, would at once sink to the center, devouring our world in the process. This would certainly be the ultimate in doomsday weapons!

Let us now imagine the career of a relativistic spacecraft which, after a journey from Earth, has succeeded in reaching a naturally-occurring rotating black hole. Having crossed the double event horizon, and taking the previously described navigational steps to avoid the destructive singularity, the crew is now on a course to another part of the universe (perhaps even to an entirely different universe, but that is the theme of the next chapter). That crew could also be on a journey into time. We remember the massive distortion of space-time within the black hole, invalidating physical laws as we understand them. And since nothing can escape from a black hole, a journey into the past can hardly lead to a violation of causality. I am fully aware that I have stressed this point already, but I feel it must be emphasized again. Such travel into the past leads only into *another* part of the universe. There is no way in which it can lead back to that part of the universe from which the spacecraft departed. Presumably, therefore, there *could* be two Miss Brights, but separated by inconceivable distance, they could never meet. Perhaps that would be as well.

Having made that point, let's look at what, at first glance, may seem an exception. So far we have maintained that once a spacecraft enters a black hole it can never return, being either crushed by the singularity or spewed out in another remote part of the universe probably in another time. Figure 24, which is an example of a standard form of diagram used by relativists, seems to contradict this premise inasmuch as it shows a spacecraft arriving back on Earth in the remote past. This does not really represent a causality contradiciton, for it has arrived back in a far-removed point in *space-time.* As you can see from the diagram, the journey is more nearly parallel to the time axis and is therefore "time-like," and never has the velocity of light been exceeded. This journey comes into the realm of the possible.

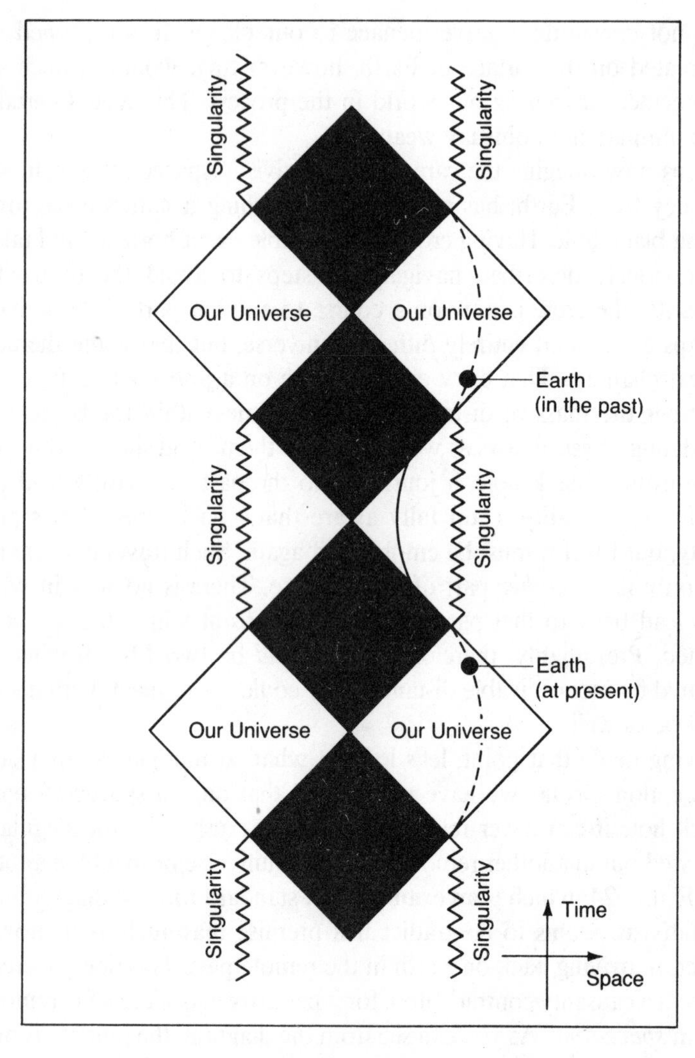

Figure 24

This standard diagram shows how a spacecraft may be brought to that point in space from which it departed, but only in the remote *past.*

This ship and its occupants could find themselves millions of years in the past. Since the Sun came into existence only some 5,000 million years ago, the time travelers might find that the Sun and the Solar System had not yet been born or, if they had, the Earth might still be too wild and primeval for even temporary occupancy. This resurrects the question about the life span of a black hole/Einstein-Rosen bridge. A small black hole mightn't exist long enough to reach or profitably employ. Table 2 relates that duration to the mass of the object.

Mass of Black Hole (tons)	Duration of Black Hole (years)
10^6	1
10^9	10^7
10^{12}	In excess of the present age of the universe

Table 2.

Relationship between masses and durations of black holes.

The black hole in the Cygnus X-1 system is believed to be six to ten times more massive than the Sun and to have a theoretical lifetime of 10^{67} years. The present age of the universe is between 10^{10} and 10^{11} years, so a fresh problem is now raised, viz., could our universe exist for more than 10^{67} years? Only if it were an "open" universe—one that continues to expand indefinitely. The prevailing belief is that the universe is a pulsating entity. The "big bang" that is thought to have marked its birth took place some 10^{10} to 10^{11} years ago. It is presently expanding rapidly, but this expansion will slow down until at 3.5×10^{10} years from now the process will terminate and contraction begin. At that stage the universe will be of maximum dimensions (5.2×10^{10} light-years). Thereafter contraction will continue until, at 9×10^{10} years from now the primeval "fireball" will be reconstituted and the next cycle of the pulsating universe will be ready to start. Thus a 10-solar-mass black hole such as Cygnus X-1, though it may have a *potential* life of 10^{67} years, will never achieve that potential, being swept up with all the other galactic and stellar "junk" of a contracting universe.

It is only fair to add that recently several sources have shed some fresh doubt on the eventual collapse of the universe preparatory to a fresh cycle of "big bang" and subsequent expansion. Might it not just continue to expand indefinitely?

The gravitational self-attraction of the universe is dependent on its density. If this is to reverse the expansion (and initiate contraction), that density must be greater than some "critical" density that would ultimately just balance the gravitational self-attraction with the force expanding the universe. This critical density works out to approximately 2×10^{-29} gms/cc. The observed density, however, gives a figure of 2×10^{-31} gms/cc, i.e., only one percent of the critical density. If these results can be accepted at face value, the universe would appear to be "open" by a wide margin, its expansion an *ongoing* process. In the context of time travel this means that black hole "time machines" would have a very much longer existence, though to the human race this could be of only academic interest. Somehow it seems hard to believe that humanity could exist for as long as 10^{67} years. In any case our planet will be consumed by the Sun in 5×10^{9} years. On the other hand, the bulk of cosmic mass probably is in a form we cannot see. Such dark mass, it is reckoned, might be 10 to 100 times that of the bright mass. This would be sufficient to "close" the universe, and this is now the more widely accepted doctrine.

Entry by a spacecraft into a black hole/Einstein-Rosen bridge/white hole system could prove the cosmological equivalent of the proverbial "blind date" since its occupants could well be taking "potluck" (to mix a metaphor) as to where and when their journey might terminate. The eventual destination might be:

(a) in *another* part of the universe in the *past;*
(b) in *another* part of the universe in the *future;*

(c) in the *same* part of the universe but in *another time* and (as we shall be seeing in the next chapter);
(d) in a *parallel* universe in the *present.*

It is clearly impossible for us to ascertain what control, if any, could be exercised. It has been suggested that crossing the event horizon in a direction contrary to that of the hole's rotation *might* be the key to whether a spacecraft would emerge in the past or future relative to Earth. This idea would seem to have more merit if such a contra-flow entry didn't raise again the fearful specter of being sucked toward the singularity. We simply are not in a position to judge how the laws of physics are changed or to what extent they are negated within a black hole system. About all that can be said with any degree of assurance is that beyond the event horizon the roles of time and space are mysteriously reversed so that those features we normally associate with time now refer to distance, and vice versa. In trying to wrap our perceptions around such notions *mind*-warp is an avocational hazard.

OTHER UNIVERSES— SIDEWAYS INTO TIME

As we have seen, journeying into the past or into the future presents many peculiar anomalies and paradoxes. Having negotiated some of these shoals, an entirely different proposition beckons. The words "time travel" by their nature conjure up visions of going backward into the past or forward into that great unknown we call the future. It therefore may come as something of a surprise to learn that there exists a possible third alternative—travel *sideways* into time. This implies neither past nor future, so you may want to object that the only other possibility is the present. And since we already exist in this, there is nothing very remarkable about that. But what if we can reach *parallel presents.*

In Chapter VI we touched upon the possibility of time being a branched entity rather than a single, continuous stream. The consequences, should this be so, would be that each and every one of us might one day reach (or have already reached and passed) this strange parting of the ways without the slightest suspicion of anything untoward having taken place. Once beyond this "time junction" we could be living blissfully in a different time, although it still would be the present. As I write these words could it be that an identical me in a parallel but entirely disparate time stream is doing something totally different amid similar surroundings though the hour and calendar date read the same?

I would be the last to claim that this is easy to conceive or to accept. Indeed it sounds like science fiction at its most improbable. As a start let us imagine a strip of 35mm photographic film containing six individual frames. On each of these frames a person is performing a particular action—e.g., walking along a quiet country road in *one* time stream. Next let us imagine *three* such strips of film, each also having six frames, and that these strips are arranged parallel to one another. These represent three alternative time streams. On the other two strips the same person may be reading and swimming respectively. In other words, on each strip that person is performing a different action at the *same point in time.* Is this really the *same* individual? The idea seems fantastic.

So far we have thought in terms of time branching from one stream into several parallel streams. Such a model is convenient as an introduction to the notion, but it is likely to be more useful as a conceptual aid than it is to mirror reality. Perhaps parallel time streams have always existed and always will, whereas the idea of one branching into several strains our credibility. So, instead, we turn once again to cosmology and think rather of parallel *universes*—replicas of our own existing "elsewhere" in the present.

Parallel universes are seen as alternative versions of the known universe, and over the years several science fiction writers have made use of the theme. At first regarded with widespread disdain, the concept has earned itself some little favor among at least a few reputable cosmologists. It has often been claimed that the universe is not just strange but a great deal stranger than we can possibly imagine. This is especially true when we come to regard the Universe as some peculiar entity embodying *many* universes. If these really exist, an answer may well have been found to some of the dilemmas that have troubled our investigation. Noted physicist Dr. John Sarfatti sums this up well: "We avoid the known paradoxes of time travel because of the many possible universes. A time traveler will probably return to a universe that is different in form but very similar to the universe from which he started. These different universes could differ in very subtle ways so that unless the time traveler is extremely observant he may not even realize that he has returned to a *different* universe."

To what extent would these other universes be replicas of our own? Would the stellar constellations show the same configurations? Did history on each Earth duplicate what *our* Earth has experienced? Perhaps in another universe the Axis powers proved victorious in World War II. Such an eventuality would certainly have rendered the years from 1945 onward very different. In such an event a time traveler would certainly be aware of more than just the subtle differences proposed by Dr. Sarfatti.

In our analogy we visualized, for the sake of simplicity, only three strips of film, representing three parallel universes. In fact we should have visualized many more, just how many we cannot say. The nearest or adjoining layers of parallel-time universes would, it is thought, contain an Earth and a Solar System so similar to the one in which we exist that, were it feasible to proceed from one to another, the differences between the two would be so infinitesimally slight that, like as not, we would be completely unaware of any change. This accords with Dr. Sarfatti's vision. It is only if such a transition were to take place over *many* layers that we would realize something highly odd had occurred. Thus, if we were walking along that country road mentioned earlier, a "jump" to the *adjoining* "layer" or universe would find us doing more or less the same thing.

In Chapter VII (Figure 19) we saw how a black hole could lead into another part of our galaxy. Figure 20 took the idea considerably further by showing how this could lead into a parallel universe. We must always bear in mind that a singularity is the focus of entry and exit points of something *beyond* space-time projecting itself *into* space-time. That "destination" space-time could and probably would be very remote from the region of space-time a spacecraft leaves when it crosses the event horizon. The actual choice of "routes" back into the universe from which the ship had departed could be many and varied. If there really does exist a stack of parallel universes, it is at best unlikely that the would-be return route the ship follows would be the one actually leading into the universe from which it had departed. The more parallel universes that exist, the less, presumably, would be the chances of finding the right one. This being so, our thoughts concerning the geometry of black holes may have to be considerably modified. We'll come back to this shortly.

Remember those four hypothetical spacecraft in Chapter VIII? Now let us see what might happen to a fifth spacecraft, E, which, like its counterpart D, crosses both outer and inner event horizons, avoids being crushed by the singularity, and emerges safely into normal space-time—only this is the space-time of a *parallel* universe. Its occupants have gone neither forward nor backward in time but remain *in the present* within *another* universe. Since there could be many routes, each leading into a different parallel universe, it is beginning to appear as if ship D and its occupants probably were lucky to emerge in the past or future of our *own* universe. They emerged not only "elsewhere" but "elsewhen." And whether this is past or future probably depends on the "route" followed. The occupants of ship E, however, could now exist in an *alternative present.*

We will now leave these spacecraft and their imaginary journeys and return to our earlier analogy involving the film strips. Remember, however, that this is *only* an analogy and that entry into parallel universes could be achieved only by spacecraft with crews somehow able to select routes within black hole/Einstein-Rosen bridge systems.

Earlier we said that could a jump be made in a "stack" of universes, from ours to one *adjoining* on "either side," we would find ourselves still walking along that same country road. Any difference would be infinitesimally slight, perhaps merely a single step forward or backward. We probably wouldn't notice, or perhaps we'd just think we had stumbled. If, on the other hand, this odd type of "quantum" jump (made in some mysterious way without recourse to spacecraft and black holes) resulted in our finding ourselves seated many miles along that same road this would be an indication that it had taken us not to a universe adjoining ours in the "pile," but to one further removed. Were the transition of considerable extent up or down the pile, we conceivably could find ourselves doing something totally different. If we were capable of recalling what we had been doing an instant before, there can be little doubt that immense confusion would reign in our minds. Chances are, of course, that we would have no recollection whatsoever of having walked down that road in our own universe.

This is very much a philosophical approach to the problem and really should not be regarded as anything more. The analogy of somehow "jumping" from one universe to another almost at will could lead to "Miss Bright"-type problems if facsimiles of ourselves exist in each universe, for presumably we might meet ourselves!

In his 1978 book *The Runaway Universe,* cosmologist Paul Davies opens his own window onto these other universes:

> Contained in this bizarre reasoning is the expectation that after a colossal number of different universes were produced in this way at random, the time would certainly come when the next fluctuation would produce an almost identical copy of our universe that we see now complete with the Sun, Earth, Empire State Building—and the reader! Of course before a world indistinguishable from our own had been recreated, an almost limitless number of "near misses" would first occur, some without the Empire State Building, even more without Africa, and many more without the Earth at all!

In Davies' approach, universe formation is a sequential process, one universe being created after another and so on. That differs from the concept of a "stack" of coexistent universes.

But the two fundamental questions are as follows:

(1) Could we make the transition from one universe or time stream to another?
(2) *Are* there really "parallel-present" universes, of which our own is merely one?

To address these questions we must once again return to the by now familar black hole/Einstein-Rosen bridge/white hole system. These could conceivably permit instantaneous transit from our part of the galaxy to another extremely remote. Those of us who continue to perceive this mode as still restricting us to the confines of our Milky Way are merely reflecting our cosmic insularity. Even imaginable transits to other galaxies can be similarly characterized for the only universe involved is still the one in which we exist. Travel to the galaxy nearest to our own (the Great Galaxy in Andromeda, generally referred to as M31) involves a destination we know and are able to see during any clear moonless night in the fall and winter. No matter how stupendous such a journey might be, we are still thinking only in terms of our *own* universe. If parallel universes exist, there is almost certainly no way we could reach one other than via the strange physical parameters of a black hole or some other form of space-time warp involving a unique type of cosmic geometry. Parallel universes by definition must exist in the present. They are contemporary entities. We cannot therefore give a definitive answer to question 1. We can say only that, due to the queer distortions of space-time within a black hole, such a transition might be feasible.

Neither is a conclusive answer to question 2 possible. All we can say is that current theory does tend to suggest the existence of parallel universes. It is hardly surprising that our supposedly rational minds shrink from such a fantastic idea. Whether it is any more fantastic than a nearly instantaneous journey through space and time via a black hole is a moot point. Today, however, as the twentieth century draws gradually to a close, the possibility appears to be gaining a measure of credence. All the infinite possibilities relevant to the future of our planet could be equally valid in a "stack" of universes just as real as the universe we know, but separated from it by some strange and indefinable cosmic gulf.

Now let us return to spacecraft E which, by taking the appropriate route through a black hole system (perhaps more by accident than design), ends up in a universe parallel to our own. If this were due purely to the dictates of chance, the occupants of that ship would with good reason feel perplexed.

Cosmologists have created a number of elaborate and complicated diagrams that demonstrate how a spacecraft could go through a black hole and emerge not in the past or future but in a parallel present universe. Figure 24 is one relatively simple variation on that theme. On reflection I felt that some readers might feel more at ease with something simpler. So while figure 25 may not portray the situation precisely, it may help you to come to terms with the realities of a highly unique and complex situation. In any event it must be regarded as grossly oversimplified.

The situation portrayed in figure 22 was greatly simplified in that it involved a simple black hole leading via an Einstein-Rosen bridge to a

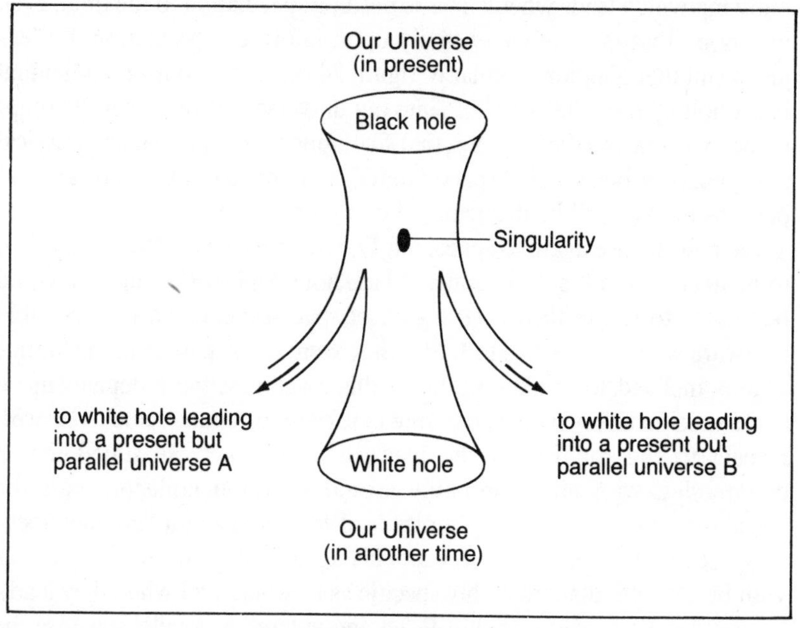

Figure 25

Hypothetical representation of a black hole system capable of leading into the past or future of our own universe as well as into parallel universes in the present. Such a configuration must be regarded as highly conjectural because of the number of parallel universes that might exist.

white hole exit. So long as the spacecraft proves successful in avoiding the singularity, it emerges in another part of the space-time of our own universe. But if such a system is also the gateway into a number of parallel universes, it could conceivably possess a multiplicity of white hole exits. A system of this kind would thus vary considerably from the standard black hole/white hole concept. It is possible, of course, that this difficulty might be obviated by postulating differing types of black holes, e.g., those that lead only to the past or to the future within our *own* universe, and those that lead exclusively into one or more *parallel* universes. Were this so, the problem facing space/time travelers presumably would be in knowing which leads where! This is pushing speculation to the limit, even in a book that must rely so heavily upon informed speculation. So let's just recall that diagrams similar to figure 24 do tend to portray individual black hole systems that lead both into our universe *and* into others. Perhaps in the final analysis the answer is that so unique and bizarre are the physical parameters of black hole "space-tunnel" systems that our minds are and perhaps forever will be incapable of comprehending them.

Back now once again to spacecraft D, which we have already assumed to have entered a "multi-channel" black hole. Since its occupants would be unable to regain their own region of space-time, could they possibly be aware whether, relative to Earth, they were in the past or in the future? One is inclined to think not, unless this awareness were dependent on some specific factor such as entering clockwise or counterclockwise with respect to the rotation of the black hole. (Earlier we stated that a counterclockwise entry probably would occasion collision with the singularity unless "clockwise" holes lead to the past and "counterclockwise" to the future, or vice versa. This again unfortunately involves us with black holes that are highly specific as to where and when they lead).

Could the occupants of ship E, having entered a parallel universe, be aware they were still, relative to Earth, existing in the present? Much might depend on which of the parallel universes they now found themselves. Should it be one far removed from our own in the "stack," the conditions prevailing there could be so different that the occupants would simply believe they had reached a remote region in the space-time of our own universe. If, however, they had gained entry to a facsimile or near facsimile of our own, they would surely be perplexed by the fact that they had *apparently* been ejected by a black hole—something known to be impossible!

Before throwing off some of the restraints of science to happily indulge ourselves in a chapter dealing with time travel and science fiction, it's appropriate here at least to mention a concept, recently put forward by more than one reputable scientist, that renders "elementary" virtually everything we've investigated so far. Some physicists and cosmologists have postulated the existence of *no less than sixteen dimensions* (not to be confused with other *universes*). Earlier we mentioned the possibility of a fifth dimension, termed hyperspace—a notion that taxes to (and beyond) the limit the capabilities of verbal definition. The demonstrations of the possibility of so many dimensions remain highly mathematical, probably intractably so. Suffice it to say that the concept may just possibly tie in with the idea of parallel present universes. Still, can you hear the sixteen echoes, one for each dimension, as Shakespeare admonishes, "There are more things in heaven and earth, Horatio, than are dreamt of in your philosophy"?

TIME TRAVEL AND SCIENCE FICTION

Writers of science fiction have always enjoyed a distinct advantage over writers of straight science, for they are permitted to take liberties that would at once put the latter beyond the pale. Such license is perfectly reasonable so long as the liberties are not too extreme. Many years ago I read a short science-fiction story in which some odd form of intelligence actually existed on the surface of the Sun! When one considers what this heaving maelstrom of plasma at 6,000 degrees C is like, I feel the author was stretching credibility a bit far. Science fiction is, of course, meant to entertain and its capacity to do so would be severely restricted were it to be placed in a scientific straitjacket.

Space travel between planets, stars, and occasionally even galaxies has for long dominated the science-fiction scene. This is perfectly understandable. Where, after all, could an author find a better stage for a genre scenario? Space is an entity that the senses can appreciate; time travel is a much trickier literary setting and must be employed with caution. Causality is the inhibiting factor, but the most blatant violations of it still appear left, right, and center. If the story is a good and gripping one, readers may well fail to spot them. However, causality, as we have seen, is such ticklish business that a fiction writer can surely be forgiven for glossing over the peculiar paradoxes that inevitably arise.

Science fiction as a serious literary form dates from around the turn of the century. At that time two men more or less dominated the scene: Jules Verne and H. G. Wells. Both wrote fascinating tales, but Verne sometimes slipped up rather badly on matters of basic science. His *Moon Voyage* is one such story, for he contrived to have three intrepid voyagers fired from a colossal cannon sunk deep within a hill near Tampa, Florida. It is all too obvious that the effect of such tremendous and sudden acceleration would have been to reduce the three to a mass of bloody pulp and bone within a fraction of a second. Wells, a graduate scientist, got around that problem in his *First Men on the Moon* by dispensing with cannons or rockets and constructing his Moonship of a material capable of negating gravity, thereby ensuring a much more comfortable and less sanguinary journey for its occupants. Needless to say, there's no evidence even now that such material could possibly exist, but Wells manages to weave such a convincing background of pseudo-science around the project that the impossible seems eminently reasonable.

Wells' *The Time Machine* was published in 1895, and during the early 1950s it was made into a highly entertaining motion picture. I saw this film again on television fairly recently. From the start anomalies abound, but it must be borne in mind that the book was written several years prior to Einstein so that the intimate relationship between space and time was not then so fully understood. Consequently the audience is confronted by an inventor who believes in the four dimensions of length, breadth, depth, and time but realizes that only in the first three does man have freedom. In the fourth dimension, time, he is a prisoner. The machine he constructs is *purely* a time machine, capable of moving in this dimension while remaining stationary within the other three. Since he knows nothing of the continuum, it cannot be taken into account in the story.

The hero of the film version sets off just before the dawning of the century year 1900 in a peculiar-looking contraption. This is essentially a chair in front of which are located the controls and instruments, which somewhat incredibly include a calendar. Located behind him is something resembling a parabolic beam antenna that rotates rapidly. The scenic effects are excellent, and through the glass partitions above the laboratory housing the machine we see the daily and accelerating passage of the Sun across the sky.

Steadily farther into the future in his strange craft goes Wells' time traveler. Occasionally he halts the machine's passage through time to ascertain what is happening in the world. On the first such occasion he discovers that World War I is in progress. Deciding this is hardly the time to linger, he proceeds on his futuristic way. Soon above the now heavily cobwebbed glass skylight he sees the Spitfires and Messerschmitts battling it out over the skies of southern England during late 1940, and he assumes the first war must still be in progress. His next stop is in 1966 (by now the view is somewhat changed, his house having been demolished by the near miss of a German bomb in 1940), and to his considerable surprise the war is apparently still in progress. Now however, although still blissfully unaware of the fact, this is another war and this time it is nuclear. Our bewildered time traveler is totally oblivious to the screaming notes of the sirens or of the warning to take immediate shelter as "an atomic satellite is homing in on the city." The resultant violent detonation totally obliterates the city and penetrates the Earth's crust, thereby releasing a torrent of molten lava (quite a bomb even by contemporary standards). Somehow the time traveler and his quaint machine escape destruction but are entombed by the lava in a great cave which cools and solidifies. Now he must advance through many thousands of years until such time as wind, frost, and rain have weathered away his magmatic tomb, exposing him once more to Sun, clouds, and blue skies. Of course his time machine has not moved in space and thus still remains in its exact 1890s position even though his house and laboratory have long since gone. Since the time traveler still appears exactly as he did at the commencement of his perilous adventure, it is apparent that travel into the future has not aged him or brought about his natural death at the allotted time. Here once more we are made aware of another paradox concerning futuristic time travel—everyone ages and dies except the time traveler!

Eventually, in the incredibly remote year of A.D. 802,701, the traveler finds, somewhat disconcertingly, that after more than 800,000 years, Earth has reverted to a new and terrible barbarism in which a kindly, docile race known as the Eloi are ruled by cannibalistic tyrants called Morloks.

In the ensuing struggle for freedom, in which the traveler is cast in the role of leader, he becomes romantically attached to a young woman. Unfortunately, in the final stages of the struggle she becomes trapped behind great doors in the former stronghold of the Morloks, and it is with the most extreme reluctance and sorrow that he returns to his own time. Since the time machine has been moved by the Morloks he finds that, even though he is now back at the date he started, it is no longer located in his laboratory but some distance away among bushes in his garden. Only one of his contemporaries believes his strange story and, still overcome by his love for the girl of the far future, he manhandles the time machine back to its former position in the laboratory and sets off once more for the distant year of A.D. 802,701. Here the film version closes, and we are meant to assume that the pair were eventually and happily reunited.

This very brief summary will have revealed a host of causality violations. Since travel in time and travel in space *cannot* be separated, *The Time Machine* and the movie based on it must be regarded as examples of literary *and* scientific license. It probably could be argued that our knowledge of space-time is still very sketchy and that great technological civilizations at other points in the universe with vastly superior knowledge might conceivably have devised some technique whereby space and time could be separated, so rendering a form of stationary "time capsule" possible. But all the evidence to date points overwhelmingly to the inseparable nature of the two. Wells' time traveler journeyed in time but remained fixed in space. To achieve the converse, to travel in space but remain fixed in time, is something with which the mind cannot even begin to grapple.

Most science-fiction writers now understand perfectly well that there is a continuum, as is evidence by Christopher Priest's 1976 novel *The Space Machine.* The similarity of this title to that chosen by Wells eight decades before is intriguing. Priest's story illustrates well the fact that time and space are inseparable, and that a space machine must also be a time machine. In his story "Tomorrow Is Too Far," written in 1977, Priest spells out the matter very clearly: "It was not time travel in the accepted Wellsian sense where a few decades of travel into the future placed the operator in the same house and room that he had left but in an older, perhaps ruined version of the same building. The present day time traveler has to be an astronaut as well."

After the appearance of Wells' *Time Machine* there does not seem to have been much fiction based on time travel. By the late '50s and early '60s, however, more was appearing. One novel I found particularly appealing was *A World Out of Time* by Larry Niven. The principal character decides to make a journey to the center of the Milky Way, traveling at almost the velocity of light. Since the Sun and its family of planets are located about two-thirds of the way from the galactic center, the journey is extremely lengthy, so much so that even relativistic time dilation proves insufficient. Accordingly the galactic voyagers are impelled to utilize the technique of cryogenic hibernation—deep-frozen sleep. When the center of the galaxy is finally reached after a journey of almost 33,000 light-years, an immense black hole is found. Not content with having achieved such an epic voyage the leader of the expedition decides to allow the spacecraft to come under the influence of the tremendous gravitational field (we know how that is, and why). The spacecraft is hurled around with tremendous violence and then flung tangentially back toward the Sun and the Solar System. When they arrive back, the ship's occupants make a profound if highly disconcerting discovery. Time dilation, it appears, results from gravitational influence to a much greater extent than that occasioned by the near-light-velocity travel experienced on the outward journey. The overall effect is to return ship and occupants not merely a few lifetimes later, as had been anticipated, but *three million years in the future.* By then, of course, Earth is a very changed place indeed.

The author of the book upon which I'll comment next certainly availed himself of license, too. Unfortunately, since that book was written more than thirty years ago and my copy seems to have disappeared on "permanent" (forgotten!) loan to someone, I'm unable to give the title or name the author. In this highly intriguing and imaginative tale, the Sun and the Solar System are swallowed up by a black hole of truly gargantuan proportions and then flung violently around the singularity to be spewed out from an equally monstrous white hole. Earth miraculously escapes destruction, but due to time reversal it is an Earth now back in medieval times. The crew of a spacecraft that earlier had passed through the black hole system are totally baffled when they see the Solar System spread out ahead of them. At first they wonder if the space tunnel has collapsed, leaving them back at their starting point, until they notice that the stellar constellations are strangely different. No longer are they the familiar star configurations seen from the Solar System. Not surprisingly the descent of the spacecraft to Earth creates considerable alarm among the planet's inhabitants; the year is A.D. 1400. Some regard the astronauts as gods to be worshipped, others as devils to be exterminated. The occupants of the ship, already thoroughly bewildered to find their familiar planet located in this remote part of the galaxy, are even more disturbed to find it so many centuries in the past. Knowing as they do what Earth's future holds, they set about trying to alter the course of history. At this point causality should have reared its head, but the author contrives to avoid this by so ordering affairs that everything the space travelers do has merely a "negative feedback" effect, preserving the status quo. History is apparently immutable—it cannot be changed. Everything is predestined. When the notorious Black Death (bubonic plague) breaks out, the crew try to use their limited but highly sophisticated medical facilities to check its spread. But, overwhelmed by the scope and rapid onset of the disease and with two of their number dead as a consequence, the ship resumes its journey in space in search of a new, more hospitable world. They leave Sun, Earth, and Solar System to their fate and future in the wrong part of the galaxy and in the wrong time. Just how they eventually regained their former correct position was understandably not explained.

Time Trap by Keith Lawson, published in 1970, is worthy of mention, for the author describes lucidly the consequences of a mysterious disturbance in the normal flow of time. We read of an ancient galleon, fully-crewed by Spaniards of the sixteenth century, being taken in tow off the Florida coast near Tampa by the present-day U.S. Coast Guard; President Lincoln is both seen and photographed in a modern Tunisian bazaar, and, as though all this were not enough, many persons find themselves living inexplicably through the same day over and over again. All these things are caused by what Lawson terms an "aperture," entry into which invariably alters a person's position both in place and time and, so far as the latter is concerned, this may be either past or present. Very odd things happen. Flowers cut and put in a vase the day before are found to be growing once again on the plants from which they were cut, persons killed the previous day show up alive, well, and unharmed on the following day—or what seems to be the following day. All attempts to return to the present are continually frustrated, for every move through the glowing aperture results only in another space and time shift. Apparently a race known as the Rhox are planning to invade Earth, although it is no ordinary invasion that is planned. This is to be an *invasion from time,* with the Rhox planning to occupy all ages simultaneously! When it appears that the hero of the story is likely to foil their plans, the Rhox threaten to open the "time-locks" so that members of each age will immediately swarm into all the others. Diplodoci will graze in New York's Central Park, early Christian martyrs will mingle with drug addicts in the Jurassic age, hordes of screaming Sioux will gallop through the suburbs of modern American cities, Pharaoh and the present ruler of Egypt will come face to face in modern Cairo. There is practically no limit to the host of bizarre possibilities. Fortunately, as might be expected, all comes right in the end. The author does not, however, go into detail as to how the mysterious time aperture is created and merely insinuates that the space-time continuum has been "disturbed."

No brief resumé of time travel as seen through the eyes of science-fiction writers could be complete without some reference to *Guardians of Time* by Poul Anderson, published in 1960. This tale is set 19,000 years in the future by which time the secret of time travel has not only been discovered but fully utilized. Time machines shuttle to and fro up and down the centuries as regularly as do the subway trains of a contemporary great city. Causality is recognized and dealt with by an organization known as the "Time Patrol." The essential function of this organization is to prevent interference with the past, so preventing the course of history being changed.

Guardians of Time is in fact an anthology of four related stories all of which were originally published in *The Magazine of Fantasy and Science Fiction* between 1955 and 1960. Just how this ability to travel at will into past or future has been achieved is not explained, but, clearly, space travel is not involved. To all intents and purposes the shuttling time machines are essentially greatly updated facsimiles of the one dreamed up by H. G. Wells. The real fascination of these stories by Poul Anderson lies in the recognition of causality and the elaborate measures taken to prevent or neutralize actions that would alter the predestined course of history.

Since black holes have been a predominant feature in this book it is interesting to see how science-fiction writers have treated them with respect to time travel. *The Well of Time* by John Light, published in 1981, is a good example. The basic theme of the story is that of a peaceful people who inhabit a planet orbiting a blue star. The astronomy involved might seem just a little dubious since blue stars are exceedingly hot bodies generally having a surface temperature of the order of 25,000 degrees C (by comparison that of the Sun is a mere 6,000 degrees C). Rigel, the bright star in the bottom right corner of the constellation Orion, is a typical example of this class of star. Such stars on the stellar time scale are young and therefore hardly likely to prove suitable locales for orbiting planetary systems, especially for planets sufficiently old to contain indigenous, intelligent civilizations. The surface of the planet envisaged by the author possesses a distinctly orange hue, which may seem a little odd with the bright rays of a great blue sun beating down upon it. These facts do not in the least detract from a well-written and carefully conceived story. Into this placid world an evil alien presence introduces a strange blight that threatens the destruction of all life upon it. The principal character in the story endeavors to save this threatened life as well as that of the unfortunate heroine, who has been abducted by the alien life-form to somewhere beyond the known universe. Our hero finds himself and his ship swept into the depths of a black hole that turns out to be a gateway into that other universe and time. Details of the black hole are a little vague, but as might be expected, once the hero crosses the event horizon some very odd things begin to occur.

Tales based on parallel present universes do not seem to abound. So far I have come on only one. This is *The Man Who Folded Himself* by David Gerrold, published in 1973. The story is based on the assumption that if only a *single* time stream were to exist, paradoxes would be inevitable and consequently time travel would have to be regarded as absolutely impossible. On each occasion that a change is made in the time stream by an individual, irrespective of how slight, another time stream is immediately created. So far as the individual is concerned, it represents the *only* time stream since it is now impossible to return to the first. The second stream is almost identical to the first but includes all the changes the person makes to it after his appearance in it. Eventually the second time stream is changed for a third.

Causality frequently gets short shrift or is blithely ignored. As a schoolboy in the summer of 1939 I read a most enjoyable short novel by Eando Binder (E. and O. Binder). Half a century later I am a little hazy about the title, but I believe it was *The Edge of Beyond* or something similar. Although the story had nothing to do with time travel, it would assuredly have been invoked. Briefly, Sun, Earth, and Solar System are in deadly peril because of the annihilation of matter. On a starry evening the only stars to be seen are restricted to one side of the sky. Somewhat oddly, astronomers, with a single exception, fail to realize the peril facing the Solar System in a few decades. Because the sky has looked this way for so long the majority believe the encroachment has stopped. This is not so, and the one astronomer who has detected the advance, along with his assistant, set out in a small spacecraft on a reconnaissance mission toward the dark region at a speed greatly in excess of light velocity. They have not gone far before they discover the presence of a young female stowaway in the form of the attractive Alora Crodell. In this story hyperoptical velocities make no difference to the time aspect, for the three arrive back on Earth eventually with the proof they sought, but it is not an Earth of the past or the future but one that has aged only to the extent the travelers have. By ignoring relativistic effects the authors have dodged the awkward causality problems that travel at many times the velocity of light would certainly have produced. Despite this flagrant disregard for relativistic laws the story was thoroughly enjoyable and entertaining which, no doubt, is why I still remember the details after so many years.

Earlier I quoted science-fiction writer Harry Harrison. His novel *Rebel in Time* is one of the most intriguing time travel works I have yet encountered. Like Moore in his *Bring the Jubilee,* Harrison selects the American Civil War as an appropriate background. Although he divorces time from the other three dimensions, as did H. G. Wells and several others, were he not to have done so, such a fascinating plot could hardly have evolved. Briefly the story centers on a present-day colonel in the U.S. security service who, despite the passing of more than a century, still retains strong Confederate sympathies and longs to reverse the outcome of the Civil War. Having gained access to a remarkable time machine, he decides to do something positive about this. Until then the machine has been able to project only inanimate objects into the future, but the wily colonel discovers that living beings can also be safely transmitted to chosen points in the past. He thereupon takes out massive loans and proceeds to purchase gold as well as a plentiful supply of pre-Civil War U.S. currency. Additionally, and most important of all, he purloins a World War II Sten submachine gun and the blueprints necessary to manufacture the weapon in bulk. He then uses the time machine to transport him back to the year 1858 and once there uses his resources to set up plants in the southern states capable of manufacturing the weapon (which was of simple design) as well as the necessary ammunition. Knowing that the Civil War would break out in 1861, he is well aware that possession of such a weapon by the Confederate armies would render them a well-nigh invincible force. For the record, the Sten had a rate of fire of ten rounds per second. The standard infantry rifle then in use by the U.S. Army had, at best, a rate of only six to seven per minute. Between the two weapons there could be no comparison. The hero of the story follows the renegade colonel back into time and, as might be expected, foils his plans—and also as a consequence avoids causality problems.

One of today's finest science-fiction writers is undoubtedly Isaac Asimov, who has, not surprisingly, featured time travel in one or two of his many books. *The End of Eternity,* published in 1955, features a member of a highly exclusive organization whose purpose is to range through past and present centuries monitoring and even altering time's many cause and effect (causality) relationships. This approach is refreshing inasmuch as most science-fiction writers elect either to ignore the paradoxes or concentrate on forms of effect that will leave the basic cause unchanged. In *The End of Eternity,* Asimov virtually goes out on a limb to mold cause and effect into a definite pattern.

Competent science-fiction handling of time travel usually requires an author of no little talent and no little familiarity with the science involved, because of causality. Still, some of the most enjoyable "time travel" stories would have lost much of their attraction had due accord been paid to causality or to relativistic effects. It was Samuel Taylor Coleridge, I believe, who was one of the first to laud the "willing suspension of disbelief" in a poetic cause. And who is there to say that time travel is not as poetic a notion as any of which mankind has ever dreamt.

EPILOGUE

"Tell 'em what you'll tell 'em," so the saying goes. "Tell 'em. Then tell 'em what you told 'em." So abide with me, please, for a few pages of summary.

This book has been a serious but essentially "popular" endeavor to come to terms in a reasonably straightforward way with the common yet mysterious fourth dimension we know as time and to ascertain, so far as is possible in the light of our contemporary technology and knowledge, if travel within this indefinable medium could somehow be achieved. A categorical answer is obviously impossible now. A number of cosmologists do seem prepared to give a measure of credence to the concept. Nevertheless it now is obvious to you that time travel will call for powers and capabilities grossly beyond anything we possess now or are likely to for the foreseeable future. Alien civilizations millennia ahead of our own may already possess such powers, but this is mere speculation. Personally I would not be surprised if at certain points within our galaxy and others the "time barrier" had been broken. Nor would I be surprised if our own kind, many centuries from now, were to achieve the same breakthrough and perhaps decide to visit the 1990s. It would surely be the greatest irony if U.F.O.s, despite all suggestions to the contrary, come not from outer space but from "outer time"! One aspect of this possibility is particularly intriguing. We are in the present, but to such visitors from Earth's future *we* would be in the past. In a sense our present is really the past, though not, of course, to us.

Probably the most fundamental of the facts we've encountered is that of the space-time continuum. Space and time cannot be separated, and although within the limits of our technology we have complete freedom to move in the three dimensions of space, we are locked like helpless prisoners in the fourth dimension, time. Within it we can move forward but only at a rate we cannot speed up or slow down—and most assuredly cannot stop. This effectively precludes all serious notions of devices that are purely *time* machines—devices that could take us backward or forward in time while remaining stationary in space. To travel into the future we also must travel in space (and this at fantastic velocities) on a round-trip. Since travel into the past calls for the use of black hole systems, we are most unlikely to be brought back to our initial starting-off point in space, and many of the strange paradoxes and dilemmas produced by causality are thus avoided. We would be in the past relative to Earth and the life we left behind. It would therefore hardly seem like travel into the past if we were unable to see the world of our birth. Causality long was regarded as proof that travel into the past was impossible. The use of a black hole/ Einstein-Rosen bridge system, if certain essential conditions are met, renders such a journey theoretically feasible, but two great impediments remain. If we take a spacecraft beyond the event horizons of a black hole and successfully avoid being crushed into zero volume by the singularity, we would appear to have little control over where and when we emerge. Perhaps when we know a great deal more about the weird geometry and strange physics of black hole "tunnels" in space-time, a choice *might* be open to us. That prospect still lies very far ahead. The other barrier is the sheer distance we must traverse to reach even the nearest black hole, wherever that may be. At present the available evidence indicates that this lies several hundreds of light-years from the Sun and the Solar System.

Traversing such immense distances would require travel at near-light velocities. Only at such speeds and their accompanying time dilation might such a journey be rendered feasible within the average human life span. At this point we see a means whereby we could be projected into the future without recourse to black holes merely by making the return journey at the same velocity though we could not reverse the process and return to our past to upset causality. But now another barrier looms up—the fantastic power requirements if we are ever to achieve velocities of such an order. No such power sources are available now, nor are they likely to be for a very long period. Nevertheless this technique has considerable and undeniable potential not only for time travel into the future but for practical interstellar travel. "Future" travel followed by a return to the present leaves us stuck on the "causality hook." If on return to the present we try to change the future by virtue of what we have learned, we cannot succeed, for, if we did, that future we have experienced would no longer be the future. Thus we are led to the concept of alternative futures resulting from alternative time streams.

Tachyons and a "tachyon drive" aside for now, we must accept that travel at or beyond the velocity of light is simply not feasible. If this velocity could be exceeded, travel into the past in normal space-time (as opposed to a black hole) would seem feasible, but at once the mathematics of the situation present us with an intractable problem—that of deriving the square root of a negative quantity; lest it be thought this is merely a mathematical paradox, we must remember the mass problem outlined in Chapter IV. From whatever angle the problem is tackled there seems no way around it. Travel beyond the velocity of light, could it be attained, would theoretically allow us to travel backward in time *within* the normal space-time continuum, but it looks as though this can never be more than an idle dream. And even were it possible, we would at once be assailed by the type of causality and paradox problem that beset Miss Bright. Travel into the past can apparently be achieved only via a black hole whereby the two identical Miss Brights can never meet. Within a black hole system, light velocity need not be exceeded to achieve a journey into the past. But how far into the past would it take us—a few score, a few hundred, or a few billion years?

Perhaps the most bizarre concept that we've visited at length is that of black holes leading into present parallel universes. This is a most difficult idea to accept, for it entirely negates the almost universally accepted concept of time as one simple ever-flowing stream; it presents us instead with the strange concept of persons existing in the present yet carrying on individual lives in *different* universes. If we accept the model of "stacked universes," our alter ego could be performing an almost identical action in the adjoining layers; in layers much farther removed such actions might be entirely different as could be our surroundings. It is hardly surprising that the conventional mind shrinks from the notion that somehow or other facsimiles of ourselves might be leading apparently normal lives in a near-replica universe.

So there the matter would appear to rest. Theoretically, time travel seems possible so long as we think in terms of space-time, but the restrictions and requirements are, to say the least, daunting. If we travel in space, the parameter of time is essential, for without it we would presumably arrive at destination B the same instant we left destination A—only neither destination, be it star or galaxy, could exist if time did not exist also. Conversely, if we wish to travel in time, in one way or another, we must also travel through space.

Even if time travel into past or future were easily and immediately possible, we would be well advised to consider the philosophical and sociological aspects. Most of us, especially if we're at least middle-aged, tend at times to desire a return to the past if that past has been a reasonably happy one. If we could press the right button and land back in the year of our choice with no recall that we had come from the future, then our lives presumably would start to repeat themselves like the rerun of a movie. (Suppose this really *is* happening now!) There would be no causality problems in these circumstances since we would remember nothing and therefore know nothing of the future and would be unable to meddle with it. In fact, so far as travel into the past is concerned, this takes care of causality in the most effective way.

Travel into the future must also be looked at along similar lines. Science-fiction travelers into the future seem never to age, but there remains the possibility that by going into the future we simply would be hastening the day of our own demise. Could we pass that date safely and smile as we gaze upon our own tombstone? This aspect seems rarely to be considered. There will come a day—none of us know when—when on the following day we will die. If that day is 30 years and 3 days from the present, and we go forward voluntarily in time by 30 years and 2 days, then tomorrow we die. It is as simple as that. And for good measure we could also be a lot more decrepit and perhaps racked by the ailments and infirmities that passing years bring. If we go forward 30 years and 4 days, then we died yesterday. If we can look at our coffin and its occupant, then there must now be two of us—one alive and one dead. The "Miss Bright" syndrome has reappeared! There is also the opposite aspect. In going back in time might we not just get progressively younger and wind up in swaddling clothes? Or as a twinkle in our father's eye?

A hundred years ago the prospect of travel to the moon would have seemed as remote a possibility as one could imagine. And I know that some readers will have found the remoteness of my speculations about man's realization of time travel not as close-at-hand as they'd have preferred. But at least one recently published piece, written jointly by professors Michael Morris, Kip Thorne, and Ulvi Yurtsever for the respected journal *Physical Review Letters* must surely invite wonder about what have seemed to be the most intransigent of problems. The paper suggests a way in which starships could one day *bypass* the law that forbids faster-than-light travel. They, too, speak of worm holes in space that would enable a space vehicle to make journeys that in conventional physics would be impossible. Adds Dr. Morris, "Travelling into the past has hitherto been considered inherently impossible because it would enable you to murder your parents before they met and you would cease ever to have existed. But this impossible paradox is avoided in our interpretation of Einstein's general theory of relativity. We suggest the existence of an infinite number of parallel universes, so that the past you travelled into would be a different past to that in which you grew up."

With a focus not completely unlike the possibilities we examined in chapter XI, the scientists draw attention to the "many worlds theory" physicist Hugh Everett proposed in 1957, which suggests that the universe is continually branching into different "states." There would be, the article adds, "another universe, just as real as our own, for example in which Mr. Dukakis had won the 1988 American presidential election."

Dr. David Deutsch, a British physicist of the Mathematical Institute at Oxford, stated recently that such branchings of reality might be detectable by a supercomputer, a machine far more sensitive than the human mind, that could perform 10,000 million calculations per second. Dr. Morris adds, "Our theory is all very speculative. A new theory might emerge that would prove it impossible. But such speculation is valuable since it encourages physicists to explore the universe in new ways."

Meanwhile, as the old newsreels always assured us, time marches on!

Appendix 1

Micro-Mu Mesons (Muons)

In Chapter VII I referred to particles known as micro-mu mesons (generally called muons) and the evidence they provide to substantiate the strange phenomenon of time dilation. For the benefit of readers who may prefer a closer and more analytical look at the subject, this appendix should prove of assistance and interest.

Time dilation will never be observed with respect to relatively large, slow-moving objects such as trains, automobiles, or aircraft. The situation in regard to muons, however, is radically different, for they are infinitesimal and extremely fast-moving. They also break up spontaneously with a half-life of 1.53×10^{-6} seconds as measured when at rest relative to an observer. A word or two is probably in order regarding the term "half-life." Should, for example, 1,024 muons be present at any one time, on average 512 will exist after 1.53×10^{-6} seconds have elapsed, 256 after $3.06\ 10^{-6}$ seconds, 128 after 4.59×10^{-6} seconds, and so on.

Thus if the muons originally numbered N, the number N_r remaining after a time t_p (p stands for proper time) will be

$$N_r = N \times (½) \frac{t_p}{T_{1/2}}$$

where $T_{1/2}$ is 1.53×10^{-6} seconds (a half-life)

Let us now consider an experiment in which we compare the number of muons from space arriving at a mouintaintop with the number arriving at sea level. It is a feature of muons that they have a wide range of velocities. It is therefore essential to arrange the receiving apparatus on the mountaintop so that it will accept only muons within a narrow velocity range. At sea level similar provision must be made. This is achieved by recourse to an arrangement of lead screens and scintillation counters. Muons having sufficient velocity to pierce the upper lead screen will have insufficient velocity to penetrate the lower as well. They will therefore terminate their passage there, decay, and be counted. Suppose the range of velocities measured at both points (mountaintop and sea level) embraces the narrow range 0.993c to 0.991c. Let us further suppose that the velocity selected is 0.992c and that the height of the mountain above sea level is 1,920 meters. Thus to pass from mountain top to sea level muons traveling at 0.992c will take

$$t_1 = \frac{1{,}920}{0.992c} \text{ seconds}$$

$$(c = 2.998 \times 10^8 \text{ kilometers/second})$$

$$= \frac{1{,}920}{0.992 \times 2.998 \times 10^8} \text{ seconds}$$

$$= 6.46 \times 10^{-6} \text{ seconds}$$

This must be regarded as an improper time (t_i) since it is derived from an improper velocity (i.e., mesured by stationary clock). If muons had clocks and the time had been measured by these, the time would have differed and would have been regarded as proper time (t_p).

We must now determine the number of muons that will have decayed in transit between mountaintop and sea level. Referring back to the equation above, we find that if N muons were present initially, the number remaining after a *proper* time interval (t_p) has elapsed will be as follows:

$$N_r = N \times (½) \frac{t_p}{T_{1/2}}$$

Where T the *proper* half-life time measured as t_p is also, in a frame in which the muons are at rest. In order to find a solution, it is first essential to convert the value of improper time measured in the second equation into *proper* time, since t_p refers to muons *at rest.*

$$t_p = t_i \sqrt{1 - \frac{v_i^2}{c^2}}$$

where t_p is proper time and t_i improper time. Thus by substitution:

$$t_p = t_i \sqrt{1 - \frac{v_i^2}{c^2}}$$

$$= 6.46 \times 10^{-6} \sqrt{1 - (0.992)^2}$$

$$= 0.815 \times 10^{-6} \text{ seconds}$$

The important point about this result is that the time as measured by a stationary observer on Earth is almost *eight* times greater than the time as measured by a muon traveling with the velocity very close to that of light.

Appendix 2

Diameters of the Navigable Apertures of Black Holes

In Chapter VIII we passed rather cursorily over the fundamental question of how the diameters of navigable apertures of black holes could vary. For readers who are keen to pursue this subject in more detail, this appendix may prove useful.

To calculate the approximate diameter of a black hole, it is first necessary to take into account the original mass of the collapsing star, remembering that if this does not exceed certain fairly well defined limits, only a white dwarf or neutron star will result.

The relevant formula is

$$\frac{4GM}{c^2}$$

where G is the gravitational constant (6.673×10^{-8}) and c, as usual, is the velocity of light. Normally this is quoted as 300,000 kilometers per second (3×10^5). For convenience here, it is advantageous to use centimeters rather than kilometers, yielding a value of 9×10^{20} centimeters per second for c^2. In Chapter VIII the formula given was

$$\frac{2GM}{c^2},$$

but this referred to the *radius* of a black hole and not to its diameter. In that instance, M represented the mass of the star, so that the relevant formula should read

$$\frac{4GM}{c^2},$$

the diameter being twice the radius.

On account of its relatively low mass, the Sun is, fortunately for us, unlikely ever to degenerate into a black hole. Its true fate, given another 5,000 million years, is to swell out into a bloated red giant, then collapse into a conventional white dwarf. For there to be any possibility of its becoming a black hole, a star must have a mass at least three times that of the Sun. Were it feasible for the Sun one far-off day to gain black hole status, the diameter of the hole would be derivable as follows. (The mass of the Sun is 1.99×10^{33} grams.)

$$\frac{4GM}{c^2} = \frac{4 \times 6.673 \times 10^{-8} \times 1.99 \times 10^{33}}{9 \times 10^{20}} \text{ centimeters}$$

$$= 5.9 \times 10^5 \text{ centimeters}$$

$$= 5.9 \text{ kilometers (3.6 miles)}$$

At a glance this might seem to be an aperture of reasonable dimensions. Unfortunately, this is only the diameter and not the *navigable* aperture. The two are *not* synonymous. In fact the navigable aperture would be very much less—a mere 60 meters! A space/time ship isn't likely to get through that. Let us now consider the minimum mass that a star must possess to become a black hole. As already stated, such a star must have a mass *at least* three times that of the Sun. Since the mass of the Sun is 1.99×10^{33} grams, the mass of such a star would be 5.97×10^{33} grams. Substituting as before to determine its diameter:

$$\frac{4GM}{c^2} = \frac{4 \times 6.673 \times 10^{-8} \times 5.97 \times 10^{33}}{9 \times 10^{20}}$$

$$= 17.7 \times 10^5 \text{ centimeters}$$

$$= 17.7 \text{ kilometers (approximately 11 miles)}$$

Once again the navigable aperture is considerably less, only 180 meters (about 550 feet or a tenth of a mile), which tends to make one wonder whether a black hole created by such a star would really be of much use to a space/time ship.

The table below shows the diameters and navigable apertures of stars having masses up to 10 times that of the Sun. The calculations are exactly as above. It will be noted that the navigable aperture is only about one-hundredth that of the diameter.

So even with stars having masses 10 times that of the Sun, the navigable aperture seems perilously low. If transit through black holes is ever to be a practical proposition, it may well be mandatory to find holes resulting from collapsed stars whose original mass considerably exceeds this value. It might therefore prove informative to ascertain the diameter and navigable aperture that would result if the most massive star known were to become a black hole. This star, called HD 47129, lies in the constellation Monoceros. This is a binary system in which each component has a mass 55 times that of the Sun. Once again substituting in the equation:

Table 3

Solar Mass (Sun = 1)	Diameter (kms)	Diameter (miles)	Navigable Aperture Diam. (meters)
1	5.9	3.7	60
3	17.7	11.0	180
4	23.6	14.6	240
5	29.6	18.4	300
6	35.6	22.1	360
7	41.5	25.7	420
8	47.4	29.4	480
9	53.4	33.1	540
10	59.3	36.8	600

$$\frac{4GM}{c^2} = \frac{4 \times 6.673 \times 10^{-8} \times 55 \times 1.99 \times 10^{33}}{9 \times 10^{20}}$$

$$= 324.60 \times 10^{5} \text{ centimeters}$$

$$= 324.60 \text{ kilometers (203 miles)}$$

Thus the navigable aperture is 3.246 meters or approximately 2 miles. Such an aperture could be considered adequate though hardly generous.

The primary obstacle is, of course, the distance separating black holes from the Solar System. Perhaps if time dilation, involving as it does velocities close to that of light, can somehow and sometime be attained, this problem can be obviated. There is a measure of irony here, for to gain access to a "space tunnel" to the past we would first be required to move into the future via time dilation.

It will be abundantly clear by now that bodies of terrestrial dimensions could never constitute a black hole with a navigable aperture. This can be shown if we calculate as above using Earth's mass (5.977×10^{27} grams). Substituting as before:

$$\frac{4GM}{c^2} = \frac{4 \times 6.673 \times 10^{-8} \times 5.997 \times 10^{27}}{9 \times 10^{20}}$$

$$= 17.73 \times 10^{-1} \text{ centimeters}$$

$$= 1.773 \text{ centimeters}$$

The navigable aperture should therefore be about *0.2 mms.* This must surely put a damper on thinking optimistically in terms of spacecraft one day being able to produce a "space tunnel" of their own at will. Perhaps somewhere in the galaxy a few highly advanced alien civilizations have already achieved this—perhaps they know what to eat or drink to shrink themselves small enough to squeeze through such a tiny opening. Should we ever have the opportunity to meet these talented creatures, however, need we be surprised if they resemble Alice . . . or the White Rabbit!

Bibliography

Albers, D. "The Meaning of Curved Space." *Mercury, 4,* (July/August 1975), pp. 16-19.

Anderson, P. *The Guardians of Time.* London: White Lion Publications, 1961; New York: Tor Books, 1988 (paper).

Asimov, I. *The End of Eternity.* London: Granada Books, 1959; New York: Del Rey/Ballantine Books, 1984 (paper).

______. "The Ultimate Speed Limit," *Saturday Review,* July 8, 1962, pp. 53-56.

Atkinson, R.J.C. *Stonehenge.* London: Hamish Hamilton, 1956.

Balfour, M.. *Stonehenge and Its Mysteries.* New York: Scribner, 1980.

Berry, A. *The Iron Sun.* London: Jonathan Cape, 1977; New York: Dutton, 1977.

Bilanuick, D.M. "Particles Beyond the Light Barrier," *Physics Today,* May 1969, p. 43.

Birch, P. "Is Faster Than Light Travel Causally Possible?" *JBIS, 37,* 3 (March 1984), pp. 117-173.

Bolton, C.T. "Dimensions of the Binary System Cygnus X-1," *Nature, 240* (1972), p. 124.

Bonner, W. *The Mystery of the Expanding Universe.* New York: Macmillan, 1964.

Borel, E. *Space and Time.* New York: Dover Publications, 1970.

Callahan, J. "The Curvature of Space in a Finite Universe," *Scientific American,* August 1976, pp. 90-99.

Cleaver, A.V. "The Phenomenon of Time Dilation," *JBIS, 12,* 9 (September 1970), pp. 378-379, 388.

Cleugh, M.F. *Time.* London: Methuen, 1937.

Corwin, M. et al. "Discovering the Expanding Universe," *Astronomy, 13,* 2 (February 1985), pp. 18-23.

Cowley, A.P. et al. "A Black Hole in Neighbour Galaxy," *Mercury, 13,* 4, (July/August 1984), pp. 106-107.

Davies, P. *The Edge of Infinity.* New York: Simon and Schuster, 1981.

______. *Space and Time in the Modern Universe.* Cambridge: Cambridge University Press, 1977.

______. *The Runaway Universe.* London: Dent, 1978; New York: Penguin, 1980.

De Peat, F. "Black Holes and Temporal Ordering," *Nature, 239,* (1972), p. 387.

Disney, M. *The Hidden Universe.* London: Dent, 1985.

Drever, R.W.P. "Gravitational Wave Astronomy," *JRAS, 18* (1977), pp. 9-27.

Dunne, J.W. *An Experiment with Time.* London: Faber, 1939 and 1958.

Durell, C.V. *Readable Relativity.* New York: Harper & Row, 1966.

Eddington, A. *The Expanding Universe.* Cambridge and New York: Cambridge University Press, 1933; Cambridge Science Classics Series (paper), 1988.

Einstein, A. "On the Generalized Theory of Gravitation," *Scientific American,* April, 1950, pp. 13-17.

Feinberg, G. "Possibility of Faster Than Light Travel," *Physics Review,* 1970, p. 1089.

Feynman, R.P. *Feynman Lectures on Physics,* 3 vols. Reading, Mass.: Addison-Wesley, 1963-1965.

Forward, R.L. "Antimatter Propulsion," *JBIS, 35,* 9 (September 1982), pp. 391-395.

Froning, H.D., *The Challenge of Interstellar Flight.* American Institute of Aeronautics and Astronautics, November 1984.

Gale, R.M. *The Language of Time.* London: Routledge and Kegan Paul, 1969.

Gamov, G. "Gravity," *Scientific American,* March 1961, pp. 94-100.

Gardner, M. *Relativity for the Million.* New York: Macmillan, 1962.

______. "Can Time Go Back?" *Scientific American,* January 1967, pp. 98-102.

Gauqueln, M. *The Cosmic Clocks.* London: Faber, 1969; San Diego, Calif.: ACS Publications, 1982 (rev. ed., paper).

Gerrold, D. *The Man Who Folded Himself.* London: Faber, 1973; Mattituck, N.Y.: Amereon Ltd., 1976.

Gibbons, G.W. "On Lowering a Rope into a Black Hole," *Nature, 240,* 77, 1972.

Goldsmith, D. "When Time Slows Down," *Mercury, 4,* May/June 1975, p. 2.

Greenstein, G. *Frozen Star.* New York: Freundlich Books, 1984.

Gribben, J. *Spacewarps.* New York: Delacorte, 1983.

______. *Time Warps.* London: J.M. Dent, 1978; New York: Dell, 1980.

______. *White Holes.* London: Paladin Books, 1977.

Harrison, H. *Explanations of the Marvellous.* London: Fontana Books, 1978.

______. *Rebel in Time.* London: Granada Books, 1985; New York: Tor Books, 1989 (paper).

______. *The Technicolour Time Machine.* London: Orbit Books, 1976; New York: Tor Books, 1985 (paper).

Hawking, S.W. *A Brief History of Time.* New York: Bantam Books, 1988.

______. "Quantum Mechanics of Black Holes," *Scientific American,* January 1977, pp. 34-40.

______ and G.F. Ellis. *The Large-Scale Structure of Space-Time.* Cambridge and New York: Cambridge University Press, 1973.

Hawkins, G.S. *Beyond Stonehenge.* London: Hutchinson, 1973; New York: Harper & Row, 1973.

______. *Stonehenge Decoded.* London: Souvenir Press, 1966; New York: Dell, 1978.

Heggie, D.C. "Megalithic Astronomy—Fact or Fiction," *JRAS, 18* (1977), pp. 450-458.

Hinckfuss, I. *The Existence of Space and Time.* Oxford: Oxford University Press, 1975.

Hodgson, S.W. *Time and Space.* London: Longman, 1965.

Hoyle, F. *From Stonehenge to Modern Cosmology.* San Francisco: Freeman, 1972.

———. *The Nature of the Universe.* Oxford: Blackwell, 1960.

Hubble, E. *The Realm of the Nebulae.* New Haven: Yale University Press, 1982.

Israel, W. "Gravitation Collapse and Causality," *Physics Review, 153* (1967), p. 1388.

Jones, R.T. "Relativistic Kinematics for Motions Faster Than Light," *JBIS, 35,* 11 (November 1982), pp. 509-514.

Kaufmann, W.J. *Cosmic Frontiers of General Relativity.* Boston: Little, Brown, 1977.

———. "Pathways Through the Universe," *Mercury, 4,* 3 (May/June 1975), pp. 26-33.

———. "Primordial Black Holes," *Mercury, 5,* 3 (May/June 1976), p. 8.

———. *Relativity and Cosmology.* New York: Harper & Row, 1973.

———. "Supermassive Black Holes," *Mercury, 7,* 5 (September/October, 1978), pp. 97-105.

———. "Traveling Near the Speed of Light," *Mercury, 5,* 1 (January/February, 1976), pp. 4-8.

Krzywoblocki, M.J. "Clock Paradox," *Astronautics and Aeronautics, 7,* 3 (March 1969), p. 18.

Lake, K. "White Holes," *Nature, 272* (1978), p. 599.

Lawden, D.F. "Phenomenon of Time Dilation," *Spaceflight, 12,* 4 (April 1970), pp. 178-179, 183.

Lawmer, K. *Time Trap.* London: Robert Hale, 1976.

Lawrence, J.K. "Future History of the Universe," *Mercury, 7,* 6 (November/December, 1978), pp. 132-138.

Layzer, D. "The Arrow of Time," *Scientific American,* December 1975, pp. 56-69.

Light, J. *The Well of Time.* London: Robert Hale, 1981.

Lilley, S. *Discovering Relativity for Yourself.* Cambridge: Cambridge University Press, 1981.

Macvey, J.W. "Creation's Dawn," *Spaceflight,* November 1966, pp. 378-412.

Marder, J. *Time and the Space Traveller.* London: Allen and Unwin, 1971.

Moore, W. *Bring the Jubilee.* New York: Avon Books, 1976.

Morgan, J.W. "Super-relativistic Interstellar Flight," *Spaceflight,* July 1973, p. 252.

Morris, M.S., Thorne, K.S., and Yurtsever, U. "Wormholes, Time Machines, and the Weak Energy Condition," *Physical Review Letters.* New York: American Physical Society, vol. 61, no. 13 (September 26, 1988), pp. 1446-1449.

Murdoch, H.S. "Recession Velocities Greater Than Light," *JRAS, 18* (1977), pp. 242-247.

Narlikar, J. *The Structure of the Universe.* Oxford and New York: Oxford University Press, 1977.

Newton, R.G. "Causality Effects of Particles That Travel Faster Than Light," *Physical Review, 162,* 5 (October 1967), p. 1274.

______. "Particles That Travel Faster Than Light," *Science, 167* (March 1970), pp. 1569-1574.

Overseth, O.E. "Experiments in Time Reversal," *Scientific American,* October 1969, pp. 88-94.

Pathra, R.K. "The Universe As a Black Hole," *Nature, 240* (1972), p. 298.

Penrose, R. "Space-Time Singularities," *Physics Review, 14* (1965), p. 57.

Peters, P. "Black Holes," *American Scientist,* September/October 1974.

Piini, E. *America's Stonehenge.* Redwood City, Calif.: Sarsen Press, 1980.

Powell, C. "Heating and Drag at Relativistic Speeds," *JBIS, 28* (1975), pp. 546-552.

Priest, C. *The Space Machine.* London: Faber and Faber, 1976.

Reddy, F. "The Universe Bit by Bit," *Astronomy, 13,* 1 (January 1985), pp. 6-17.

Rees, M.J. "Our Universe and Others," *JRAS, 22* (1981), pp. 108-124.

Ridley, B.K. *Time, Space and Things,* Cambridge: Cambridge University Press, 1984.

Rosenfeld, A. "A 3 Million Year Trip in 55 Years," *Life, 54,* 21 (May 24, 1963), pp. 28-31.

Rothman, M.A. "Things That Go Faster Than Light," *Scientific American,* July 1960, pp. 142-8.
Ruffini, R. et al., "Introducing the Black Hole," *Physics Today,* June 1971, pp. 30-36.
Sachs, M. "A Resolution of the Clock Paradox," *Physics Today,* January 1971.
Sexl, R. & H. "Curved Space-Time Near a Neutron Star," *Mercury, 9,* 2 (March/April 1980), pp. 38-39.
Shipman, H.L. *Black Holes, Quasars and the Universe.* Boston: Houghton Mifflin, 1976.
Shu, F. "Large Scale Geometry of Spacetime," *Mercury, 12,* 6, (November/December 1983), 162-171.
Sklar, L. *Space, Time and Space-Time.* Berkeley, Calif.: University of California Press, 1977 (paper).
Solomon, J. *The Structure of Space.* London: David and Charles, 1973.
Stehling, K.R. "Time Dilation in Space Travel," *Space/Aero Engineering,* May 1959, pp. 43-45.
Stimeto, R. et al. "Celestial View from a Relativistic Starship," *JBIS, 34,* 3 (March 1981), pp. 88-89.
Stolus, G.M. "Cygnus X-1: Black Hole?" *Mercury, 8,* 3 (May/June 1960), pp. 60-61.
Taylor, J. *Black Holes.* London: Souvenir Press, 1973.
———. "Particles Faster Than Light," *Science Journal,* 43, September, 1969.
Thorne, K. "The Search for Black Holes," *Scientific American,* December 1974, pp. 33-43.
Toben, R., and Fred A. Wolf. *Space-Time and Beyond.* New York: E.P. Dutton, 1974, 1982 (paper).
Von Buttler, J. *Journey to Infinity.* London: Fontana Books, 1976.
Wells, H.G. *The Time Machine.* Available in several reprint editions.
Wheeler, J. *Geometrodynamics.* New York: Academic Press, 1962.
Whitrow, G.J. *The Nature of Time.* New York: Holt, Rinehart & Winston, 1973.

———. *What Is Time?* London: Thames and Hudson, 1972.
Wick, G. "The Clock Paradox Resolved," *New Scientist, 261,* 3, (February 1973).

**JBIS* is the Journal of the British Interplanetary Society
JRAS is the Journal of the Royal Astronomical Society
Mercury is a publication of the Astronomical Society of the Pacific
Spaceflight is a publication of the British Interplanetary Society

Index